Big Data Analytics 2016

Topical Conference at the 2016 AIChE Spring Meeting and 12th Global Congress on Process Safety

Houston, Texas, USA
10-14 April 2016

ISBN: 978-1-5108-2481-2

Printed from e-media with permission by:

Curran Associates, Inc.
57 Morehouse Lane
Red Hook, NY 12571

Some format issues inherent in the e-media version may also appear in this print version.

Printed by Curran Associates, Inc. (2016)

For permission requests, please contact AIChE
at the address below.

AIChE
120 Wall Street, FL 23
New York, NY 10005-4020

Phone: (800) 242-4363
Fax: (203) 775-5177

www.aiche.org

Additional copies of this publication are available from:

Curran Associates, Inc.
57 Morehouse Lane
Red Hook, NY 12571 USA
Phone: 845-758-0400
Fax: 845-758-2633
Email: curran@proceedings.com
Web: www.proceedings.com

Big Data Analytics 2016

Topical Conference at the 2016 AIChE Spring Meeting and 12th Global Congress on Process Safety

Houston, Texas, USA
10-14 April 2016

AIChE
The Global Home of Chemical Engineers

TABLE OF CONTENTS

444849 Beyond Big Data Analytics Integrating with Cognitive Computing to Enable Smarter Manufacturing

Abstract:

The oil & gas, downstream and petrochemicals industries are facing challenges such as aging workforce with wide gaps in experience, aging asset base with more complex operations, market variability and speed of change requiring faster and better decisions at real-time, production and energy optimization, safe, secure and reliable operations for compliance.

We will discuss how the winning companies can step up to meet the above challenges by moving beyond the capabilities of Big Data Analytics by integrating Analytics with Cognitive Computing to enable real-time Cognitive Manufacturing. This will enable real-time decision management leveraging the insights and benefits from Analytics using cognitive systems such as IBM Watson that can:

- Ingest unstructured and structured data in all its forms.
- Understand unstructured data, through sensing and interaction;
- Reason with Bid Data by generating hypotheses, considering arguments, and recommendations; and
- Continuously learn from training by experts, from every interaction and from growing data.

IBM Watson has evolved from a single capability of natural language Q&A powered by few technologies such as machine learning and natural language processing (as seen in Jeopardy! in 2011) to more than 28 capabilities such as relationship extraction, personality analysis, tone analysis, concept expansion and trade-off analytics powered by 50 technologies. These capabilities have all been turned into digital services platform offered via APIs. These capabilities are provided as cognitive building blocks in in a cloud based ecosystem which anyone could build into their applications to scale and accelerate innovation.

Decision management platforms utilizing IBM Watson will meet the need for greater consistency in optimal decision making and knowledge retention by providing the capabilities such as the following: 1. Deeper human engagement, 2. Elevated expertise, 3. Cognitive products and services, 4. Cognitive processes and operations, 5. Intelligent exploration and discovery.

444732 How to Start Your Company on the Big Data Journey

Abstract:

The promise of Big Data in the chemical manufacturing industry appears to be coming to fruition. In this presentation we tackle the vexing problem of convincing your organization when and how to take the journey into Big Data. In this paper we will offer our experiences, reveal some of the challenges we met in gaining buy-in and how we overcame these obstacles.

444903 Harnessing Big Data for Smart Manufacturing: A Novel Optimization Framework

Abstract:

Manufacturing planning and management practices traditionally feature vertically optimized operational units, where enterprise segments are optimized in a vacuum, planning and scheduling rely on conservative forecasting, and reactive decision making is slow and hindered by incomplete information and inherently static infrastructure. Advanced smart manufacturing frameworks are characterized by enterprises which are fully-integrated and able to harness and optimize vital operational, business, and supply chain information in real time. Current and developing capabilities in the rapid and decentralized acquisition of data both enables and complicates the development of smart manufacturing methods. Existing and new methods that are to be applied to Big Data for smart manufacturing must be able to accept and prioritize highly variable data sources, including factory sensors, historical and real-time information, images, market and demand data, and customer-side product performance collected via the cloud. Applicable methods must also be efficient to draw rapid conclusions such as for fault detection and root cause analysis while still capturing the highly nonlinear and interconnected nature of the data inputs.

In this plenary talk, we will present and discuss optimization- and statistics-based frameworks for data-driven decision making which can identify the key components of an input high-dimensional dataset. The methodology can be applied to a variety of situations under the domain of smart manufacturing by modularly incorporating machine learning approaches, including those for nonlinear regression, classification, outlier detection, or clustering. Optimization models were formulated which cast the high-dimensional feature selection problem into implicit nonlinear feature space, resulting in minimum-maximum mixed-integer nonlinear optimization problems.. Extensive computational studies over numerous benchmark datasets show that new algorithms outperform existing state-of-the-art methods in the machine learning literature and show great promise for accurate, real-time decision making from Big Data in smart manufacturing operations.

444831 Challenges and Research Directions in Big Data

Abstract:

Big Data has opportunities for diagnostic, prognostics, and decision making. This presentation describes challenges associated with Big Data that need to be resolved for these opportunities to be fully realized. The challenges include how to best

(1) extract information from high-order tensor data structures, which arise from spatially resolved photoluminescence, spectral imaging, and color video

(2) integrate information from a variety of types, such as process flowsheet structure, causal relationships between variables, and real-time sensor data

(3) manage the effects of uncertainties, disturbances, faulty sensors, variation in operator practices, and machine drift

(4) address nonlinear dynamic operating conditions of simultaneous discrete and continuous nature such as startup, shutdown, and product changeovers

Several examples are provided where wrong conclusions or poor performance were obtained during the seemingly reasonable application of data analytics, which illustrate some common misconceptions and the potential pitfalls of applying methods naively. Some specific research directions motivated by the challenges are outlined.

Effective Relief System Evaluation through Process Safety Information Optimization

Srihari K Maganti
PSM Consultant
15995 N Barkers Landing, Suite 145, Houston, TX 77079
srihari.maganti@provenanceconsulting.com

Sara Tzaperas
Line of Service Manager, ARTS™
815 Brazos Street, Unit 204, Austin, TX 78701
sara.tzaperas@provenanceconsulting.com

Prepared for Presentation at
American Institute of Chemical Engineers
2016 Spring Meeting
11th Global Congress on Process Safety
Houston, Texas
April 10-14, 2016

Effective Relief System Evaluation through Process Safety Information Optimization

Srihari K Maganti
PSM Consultant
15995 N Barkers Landing, Suite 145, Houston, TX 77079
srihari.maganti@provenanceconsulting.com

Sara Tzaperas
Line of Service Manager, ARTS™
815 Brazos Street, Unit 204, Austin, TX 78701
sara.tzaperas@provenanceconsulting.com

Keywords: Process Safety Information, Data Repository, Data Aggregation, Data Validation

Abstract

Process Safety Information (PSI) is a vital element of Process Safety Management (PSM) per Occupational Safety and Health Administration (OSHA) CFR 1910.119 (d). An engineer needs a variety of PSI documents for relief system evaluation, such as Process and Instrumentation Diagrams (P&IDs), Process Flow Diagrams (PFDs), Heat and Material Balances (H&MBs), relief device specification sheets, equipment specification sheets, equipment mechanical drawings, manufacturer data reports, Process Hazard Analysis (PHA) documentation, Management of Change (MOC) documentation, etc.,. Obtaining accurate PSI can be a daunting task for a Relief Systems Engineer. Process safety optimization can be achieved by an efficient and reliable data repository that can save valuable engineering time. Such a repository can be built by aggregating equipment information from multiple data repositories into a single point of access. This paper describes the use of a data aggregation and validation tool and outlines other PSI needs that can optimize a relief system evaluation.

1. Introduction

Data management plays a major role in every industry. This is particularly true for the oil and gas industry where facilities often have complex workflows, requiring significant amounts of data to perform any operational activity. PSI is the foundation of these activities and must be kept complete and accurate in order to maintain a safe and operable environment [1]. As defined by OSHA CFR 1910.119 (d), PSI consists of information pertaining to process technology (such as process chemistry, process simulation, heat and material balance, PFDs, and P&IDs), process equipment (such as U1-A form, mechanical drawings, specification sheets, repair reports, and MOC documentation), and process hazards (such as Material Safety Data Sheets and PHA documentation).

In order to perform a Relief Systems Evaluation (RSE), a Relief Systems Engineer needs a variety of PSI documentation and PSI data points, such as inlet, outlet, orifice sizes, and set pressure for a relief device; maximum allowable working pressure (MAWP) and dimensions for fixed equipment; and Heat and Material Balance and process simulation. This information often resides in multiple data repositories across many departments within a facility. Obtaining accurate and up-to-date PSI can be a daunting task, and PSI discrepancies can lead to errors or misjudgment, which and pose a significant risk to the facility [2]. Accuracy, completeness, and ultimate confidence in this data are frequent challenges to a Relief Systems Engineer, so significant time is often spent tracking down the right information.

In a typical RSE, the task of obtaining accurate and up-to-date PSI is a step embedded within the evaluation process; as discrepant or missing information is discovered by the Engineer, he or she seeks out the accurate value by reviewing the available data or document sources at the facility. However, there are pitfalls to this process that allow for significant inefficiencies and inaccuracy of data. Redefining the RSE process to focus on effective data management significantly reduces cost and increases accuracy. Effective data management must focus on integration of the multi-departmental data and document sources, traceability of the required data back to the validating source, optimization of the process to gather the data, and employing a data management tool to facilitate this process. Data management is the foundation of an RSE, and an effective RSE can only be achieved through effective data management.

2. Typical PSI validation process for RSE – Case Study

As a case study, assume an RSE is to be performed for an entire unit or facility. Assume that this facility has access to historical RSEs in an electronic relief systems design database. Additionally, assume the facility has access to various electronic and/or hard copy documentation and PSI located in multiple data repositories that are maintained by separate departments. Examples of data repositories include maintenance databases such as SAP and Maximo, and inspection databases such as PCMS, Meridium, and UltraPIPE.

2.1 Typical validation process

When performing an RSE, the steps involved in the typical validation process are as follows:

2.1.1. *Initiate relief system evaluation*
The Relief Systems Engineer starts by reviewing and evaluating the previous RSE design database on a system by system basis.

2.1.2. *Discover missing and discrepant data*
As the Engineer evaluates the system, the Engineer reviews the PSI residing in the RSE design database for that system to check for accuracy and completeness. To check for accuracy, the data in the RSE design database is typically cross-referenced with the data on the P&ID. This cross reference is performed to catch any changes made since the last evaluation. Furthermore, the Engineer checks for the accuracy of the data by performing a sanity check of the data while evaluating the system, e.g., asking *is the relief valve set pressure less than or equal to the MAWP of the vessel?* Finally, the Engineer checks for completeness by determining what information required for the evaluation is not available in the RSE design database or on the P&ID.

2.1.3. Gather PSI for missing and discrepant data

Once missing or potentially discrepant information is identified, the Engineer seeks out the accurate value. The process by which this information is sought out is typically ad hoc based on where he or she thinks the information will most likely reside. For example, if a vessel MAWP is missing or discrepant, the following steps may be taken:

a. Search electronic documentation database for the manufacturer's data report (U-1A), nameplate rubbing, vendor drawing, etc.
b. Search hard copy files for paper documentation
c. Call the Inspection Department to see what relevant information resides in their database
d. Field verify to obtain data stamped on the nameplate
e. Call the vendor

2.1.4. Populate RSE design database with new PSI

Once the correct data value is obtained, the value is populated by the Engineer into the RSE design database.

2.1.5. Complete RSE

Now that all required available data is obtained, the Engineer can finish the evaluation.

2.2 *Pitfalls associated with typical validation process*

When an RSE is performed for one or two systems, the above validation process may not be unreasonable; the engineer can take the time to ensure the accuracy and completeness of the data by manually comparing and gathering appropriate PSI. However, for larger RSE efforts or even smaller efforts repeated over time, there are a number of pitfalls in this process that can cause significant time delays and reduce confidence in the data.

2.2.1. Inefficient PSI Gathering

The process to *Gather PSI for missing and discrepant data* can lead to redundant or inefficient work over the lifetime of the RSE effort. For example, for the first system, the Relief Systems Engineer may need to call the Inspection Department to ask for assistance in verifying the MAWP of a vessel. Then for the next system a few days later, the Engineer may need to call the Inspection Department again to ask for the dimensions of a different vessel. This redundant effort is a time burden on both the Relief Systems Engineer and the Inspection Department. This same inefficiency can also be applied for any step in the PSI gathering process: going to the field to obtain nameplate information on neighboring equipment, walking to the file room to review a single hard copy file each time for different equipment, etc. Furthermore, if the RSE is performed by an offsite contractor, these inefficient requests for information system by system place an unnecessary financial or time burden on the client.

Additionally, the PSI gathering process can take time and create a bottleneck for the Engineer performing the RSE. If the Engineer has to request information from a different department or the vendor, it may take days, weeks, or months to receive a response. Once the information is provided, the Engineer has to familiarize themselves with the system again before starting the evaluation. Though this may seem to be minimal, over time the delays and re-familiarization can cause significant time delays.

2.2.2. Reduced Confidence in Data

During the *Discover missing and discrepant data* step, it is the responsibility of the Relief Systems Engineer to discover the discrepant information. However, that process allows for possible oversights. For example, in the case study outlined above, the discovery of discrepant data is dependent on up-to-date and accurate P&IDs; however, if the P&ID data is lacking or if there was a change that accidentally did not get captured on the P&ID, the Engineer may not catch the inaccuracy.

Furthermore, the data that is available on the P&ID or in the RSE design database is not always easily traced back to the validation source, such as the manufacturer's data report or nameplate. Therefore, when there is a discrepancy, it is hard for the Engineer to know what to trust or where to look for the correct value.

3. Optimized PSI validation process for RSE

The pitfalls associated with validation process can be minimized by redefining the validation process. Before the Relief Systems Engineer even begins the RSE work, the PSI data validation effort can be defined, executed, and documented. Streamlining and frontloading this effort drastically increases the efficiency and accuracy of the RSE data validation process. Additionally, utilization of a PSI Verification and Traceability Software tool can facilitate this effort, allowing for increased efficiency in the process, increased accuracy and confidence in the data, and long term reusability of the effort.

3.1. Optimized Validation Process

3.1.1. Define required PSI and establish hierarchy of validating sources

Prior to starting an RSE, the Engineer can generally define the PSI data points required for the evaluation. For example, for each vessel, the Engineer will need to know the MAWP, design code, dimensions, orientation, head types, height above grade, and vacuum rating. As the Engineer performs the RSE, other data points may ultimately be required, but a predefined list can capture a majority of the required data. Similar predefined lists can be generated for each equipment type as well as for the relief valve.

Once the required PSI data points are identified, the Engineer can establish a list of sources from which each data point can be validated. For example, if a vessel MAWP is missing or discrepant, typical validating sources are the manufacturer's data report, manufacturer's nameplate, vendor drawing, specification sheet, P&ID, inspection software, and contacting the vendor. These validating sources may vary from facility to facility but can be defined by the Engineer prior to the execution of the RSE.

Now that the required data points and the recommended validating sources for each data point have been defined, a hierarchical procedure to find this information can be established. A potential hierarchy of these sources to obtain vessel MAWP may be as follows:

1. Check the electronic documentation for the manufacturer's data report (U-1A), nameplate rubbing, vendor drawing, etc.
2. Check the hard copy files for paper documentation
3. Check the nameplate on vessel in the field
4. Check the P&ID
5. Call Inspection Department to see what resides in their database
6. Call the vendor

Again, this hierarchy may vary for different facilities; for example, for a vessel MAWP, some facilities may have high confidence in their P&IDs while others would prefer to confirm with the manufacturer's nameplate. Either way, this process can be pre-defined by the Engineer.

3.1.2. Identify missing and discrepant PSI

Now that the required data points have been determined, each data point can be reviewed to identify missing and discrepant values. This may seem like a daunting task; however, this task can be streamlined by leveraging the facility's electronic data repositories. Data from each departmental repository can often be extracted in mass and then loaded to a single system for easy data comparison. Leveraging other departmental data repositories increases the accuracy of the data: rather than just having the Engineer manually compare the RSE design database to the P&IDs, this process also compares it to other departmental repositories, which is likely to identify additional discrepancies.

Employing a Verification and Traceability Software tool can make this task extensively automated. For example, Figure 1 below shows an example of data from each departmental repository uploaded to a single system. Once in the same system, the data can be cross referenced to identify discrepant and missing data: the ✖ represents discrepant values and the ⊘ represents missing values.

	Inspection Database	Drawings (P&IDs)	Maintenance Database	Relief Systems Database
Unit No	Unit 101	Unit 101	Unit 101	Unit 101
MAWP	✖ 385	✖ 385	✖ 385	✖ 305
MAWT	✖ 475	✖ 475	✖ 475	✖ 500
MDMT	⊘	n/a	n/a	n/a
Diameter	⊘	1' 11"	n/a	1.9
Length	⊘	✖ 13' 2-3/16"	n/a	✖ 3.7
Shell Material	SA-516-70	SA-516-70	n/a	n/a
Head Material	SA-516-70	SA-516-70	n/a	n/a
Design Code	S8/D1	n/a	n/a	n/a
Head Type	n/a	n/a	n/a	Elliptical
Config/Section	n/a	n/a	n/a	VV
Vacuum Rating	n/a	n/a	n/a	⊘

⊘ Missing Value

✖ Discrepant Value

Figure 1: *Integrated equipment information from different data resources using ARTS™*

3.1.3. PSI Gathering and Validating

With the potential discrepant and missing values now identified, a Data Technician can follow the pre-defined hierarchy of sources to validate these data points. Since the data comparison effort was done simultaneously, the process for gathering the PSI can be streamlined. For example, using the example hierarchy above, the Data Technician can first access the electronic database to search for the manufacturer's data report, nameplate rubbing, vendor drawing, etc. for all vessels with discrepant or missing data; this search may be successful for confidently validating 40% of the discrepant and missing values. Then for the remaining 60% of the values, the Data Technician can make one trip to the hard copy file location and search for all possible files at once; this search may be successful for an additional 20% of the values. And so forth. Eventually the gathering and validation is performed for each step in the hierarchy, after which all options for tracking down available data are exhausted.

By consolidating the validation of all discrepant and missing values into one PSI gathering and validation effort, redundant steps are minimized, reducing the overall time spent to perform this work. It also minimizes the request burden on other departments; for example, once the Data Technician reaches the "contact the Inspection Department" step in the hierarchy, the Data Technician only needs to contact them one for a single all-encompassing request, rather than multiple small requests. Furthermore, this entire effort can be performed by a Data Technician rather than an Engineer; once the Engineer defines the pre-defined lists and hierarchy, the actual gathering and validation steps can be proceduralized and executed by a lower cost resource, allowing the Engineer to focus on engineering work.

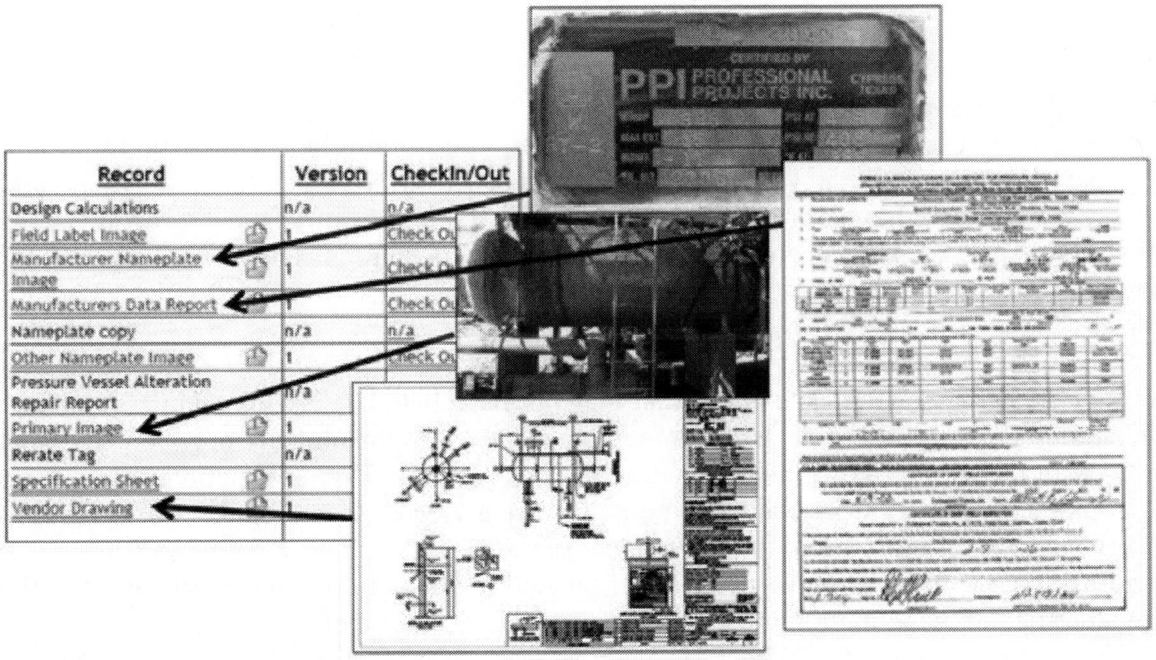

Figure 2: *Link all validating documentation with the equipment asset obtained from ARTS*^{TM}

If a Verification and Traceability Software tool is employed, this gathering and validation process can be documented for quality checking and use for future efforts. As seen in Figure 2, all of the PSI obtained during the gathering step can be linked to the equipment assets within the tool. Then, since both the data and validating documentation are all linked together into one system, the data points can be literally traced back to the validating source as seen in Figure 3. This traceability dramatically increases the Engineer's confidence in the validation process: not only can they see the value has been validated, but they can also see by which validating source it was validated. If this system is maintained over time as changes are made to the facility, then this system can be used again, *eliminating the entire data validation effort for all future RSE work.*

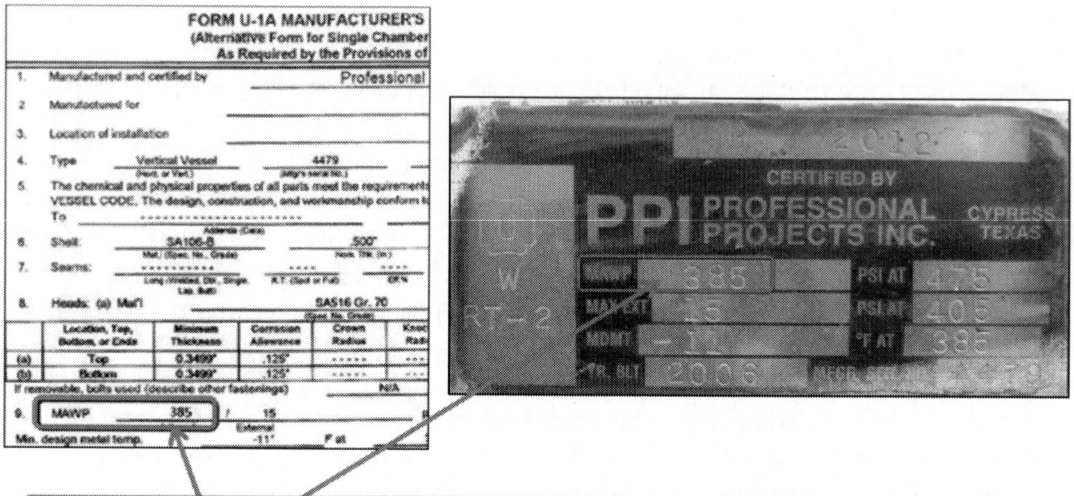

		Master	Inspection Database	Drawings (P&IDs)	Maintenance Database	Relief Systems Database
MAWP		385	385	385	385	✗ 305
MAWT		475	475	475	475	✗ 500

Figure 3: *PSI data points for traceability obtained from ARTS^{TM}*

3.1.4. Perform Relief Systems Evaluation

With the data validation process complete, the Relief Systems Engineer can begin the RSE. For each data point defined in the very beginning of this process, the Engineer knows what validating sources were reviewed to ensure accuracy of the data. If a Verification and Traceability Software tool was used, the Engineer can reference this tool for confirmation. Therefore, for each of these data points, the Engineer does not need to perform any additional PSI gathering or validation work.

Invariably, there are certain data points that may not be able to be determined through the validation process, for example, the documentation for a 70-year-old tank may be unavailable. If the result of the data gathering and validation process was a lack of information, then the Engineer knows exactly what validating sources were reviewed; knowing that all validating sources were exhausted; it is up to the Engineer to make an assumption based on best engineering judgement. Ultimately, the Engineer will not be required to perform any additional PSI gathering, allowing the Engineer to complete the effort more efficiently.

3.2. Benefits of the optimized process

For a large RSE or for smaller efforts that are repeated over time, reorganizing the overall RSE process as outlined can achieve increased efficiency, improved accuracy of the data, and therefore improved accuracy of the results of the RSE.

1. Once defined by the Relief Systems Engineer, the PSI validation effort can be performed by a Data Technician, allowing the Engineer to focus their time on performing the RSE rather than tracking down validating sources.
2. Front loading the PSI validation effort allows for a more streamlined effort with fewer trips to field, fewer trips to hard copy files, and fewer calls to other departments and vendors.
3. The Relief Systems Engineer does not begin RSE until the PSI validation effort is complete, allowing the Engineer to complete the effort more efficiently.
4. Consolidation of the data and documents into one location prior to starting the RSE allows the Engineer to easily navigate and reference the PSI.
5. Leveraging data from various departmental repositories allows for the data in the RSE design database to be cross checked for accuracy.
6. Data points are traceable back to the validating source, improving confidence in the data.

Furthermore, this improved process has benefits outside of the single RSE effort, particularly if a PSI Verification and Traceability Software is utilized.

1. Lowers the overall facility risk level by allowing discrepancies to be identified and corrected
2. Facilitates knowledge sharing between all departments
3. Provides accessibility of PSI for all departments
4. Provides validated and traceable data for future efforts
5. Improves efficiency and accuracy of long term PSI maintenance
6. Allows for a quality check of MOC process
7. Provides protection and confidence in PSI and RSE for future audits.

4. Conclusion

Redefining the RSE process with a focused effort to leverage effective data management and PSI optimization significantly reduces cost and increases accuracy. Pre-defining the PSI validation process and completing it prior to beginning the RSE effort increases the efficiency of both the validation and RSE efforts, shifts the burden of data gathering away from the Relief Systems Engineer, and reduces costs by allowing the Engineer to focus on engineer work – all while increasing accuracy. Furthermore, employing a PSI Verification and Traceability Software can significantly reduce the risk to the facility as it allows on-going cross checking of data sources and traceability between data points and validating documentation. Data management is the foundation of an RSE, and an effective RSE can only be achieved through effective data management. If done well, the value of that data validation effort can extend beyond the RSE, benefiting the entire facility on a daily basis.

5. References

[1] Mahmoud, Reza S., "Data Mining and Data Gathering in a Refinery", Proceedings of the 10th WSEAS international conference on Applied computer science, 2010.

[2] http://provenanceconsulting.com/software/arts-asset-and-records-traceability-system/

16AIChE Spring Meeting, Houston, TX

12TH GLOBAL CONGRESS ON PROCESS SAFETY

Paper 24b: Extracting Value From Plant Data

Doug White
Principal Consultant
Emerson Process Management
Houston, Texas

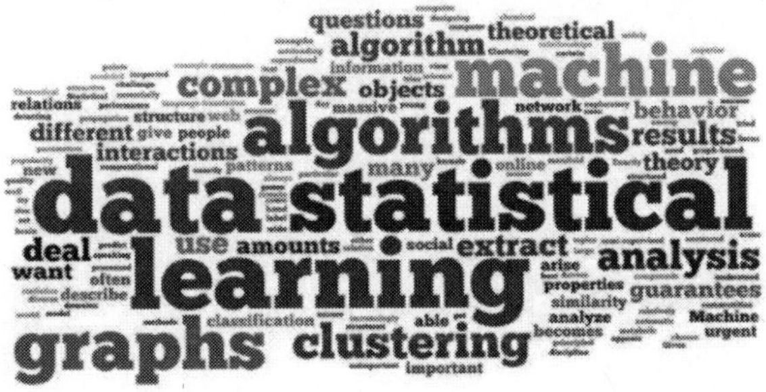

Speaker

Doug White

Principal Consultant

PlantWeb Solutions Group

Emerson Process Management

Background: Many years experience designing, justifying, installing and commissioning advanced real time computer/ automation applications in the process industries.

EMERSON. Process Management

Data is Increasing

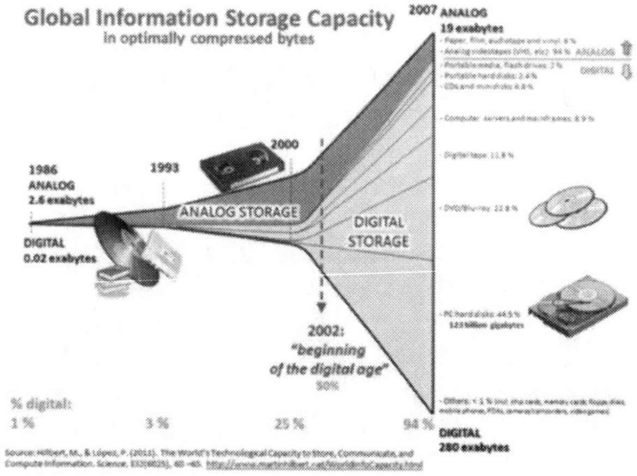

Exponential growth of global storage capacity and processing power

Data Is Increasing

- Walmart's Online Database
  - More than 250 billion rows
  - More than 20 million updates per hour
  - 3 years history available online
  - 24,000 queries per hour from 1500 users
- Internet Devices
  - Gardner group estimates 5 billion connected devices in 2015 – 25 billion by 2020
  - Plant device connectivity is increasing similarly
- Process Company
  - 10 million tags across 15 sites
  - Implementing system to provide real time access to 3 years of 1 minute sampled historical data for tags

Creating Value From Data - Predictive Analytics – Looking Forward

Looking Back

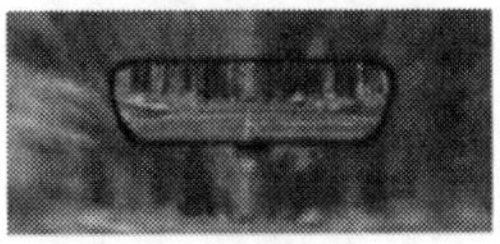

What happened and how did it compare to plans?

Gaps? Why?

What is likely to happen in the future?

Fix or Mitigate Problems

Looking Forward

Predictive analytics, in general, is the use of current and historical data in conjunction with system models to predict future trends in real-time (near).

3/27/2016

5

Where is Predictive Analytics in the "Hype" Evolution?

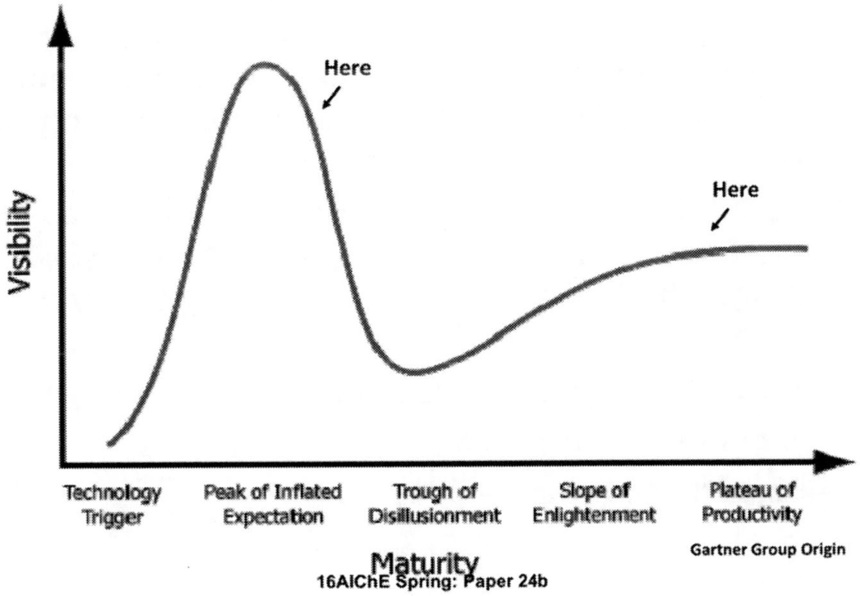

Gartner Group Origin

3/27/2016

16AIChE Spring: Paper 24b

6

Predictive Analytics - Moneyball

- In 2001, the Oakland A's (salaries $44m) lost to the New York Yankees (salaries $125m) in the divisional playoffs and then lost 3 top players to free agency.
- Billy Beane, the manager, hires Peter Brand, a new graduate, to implement statistical analysis of previous seasons to determine what it takes to win. He develops correlations that go against some of the traditional baseball ideas. Beane supports using the approach to pick new players.
- In 2002, The A's set a league record by winning 20 games in a row on the way to winning 95 games overall and making the league playoffs – and doing this again in 2003.

Predictive Analytics - Predicting Supreme Court Decisions

- Andrew Martin, a professor of political science at Washington University, developed a statistical model of supreme court decisions based on actual decisions from 1994 to 2001 (no change in justices during that period.
- He used the model to predict decisions for the 2002 term. Results were compared against a panel of distinguished legal individuals experienced with the supreme court.
- For the 68 cases in the 2002 term:
  - Model accuracy – 75%
  - Experts accuracy – 59%

Predictive Analytics – Predicting Flu Trajectory

- Google researchers used 50 million health related search queries for the period 2003 to 2008 to correlate with CDC data on the number and region for physician visits for influenza like symptoms
- Using analytics techniques they found which terms correlated most strongly with the physician visit data (correlation = 0.9)
- Using a predictive model for 2008/ 2009 flu season they were able to predict the number of visits per region about two weeks prior to the CDC report – allowing earlier response

3/27/2016 16AIChE Spring: Paper 24b 9

AFPM
American
Fuel & Petrochemical
Manufacturers

How does this apply (and not apply) to the process industries?

16AIChE Spring: Paper 24b

Plant ("Near") Real Time Data Includes Many Types

Numeric Data	Spectral Data	Video/ Images	Text
Process/ Offsites/ Utilities/ Terminal Data	Analyzers – NIR, etc	Cameras – Flares, Process, Security, Etc	Daily Plans, Notes, etc
Lab Data	Rotating Machinery – Vibration, etc.	Thermography	Maintenance - Inspection, Work Orders, etc
Accounting, KPI's, Planning, Scheduling, etc	Infrared/ Acoustic/ Ultrasonic – Event Detection	Misc	E-Mail

Typical Refinery DCS IO (Input / Output) Numbers – Continued Increase

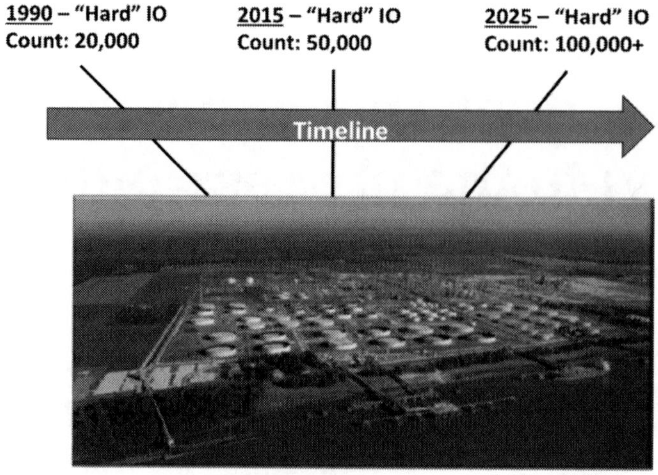

Photo Source: US DOE

Remote Access - "Internet of Things" in Plants

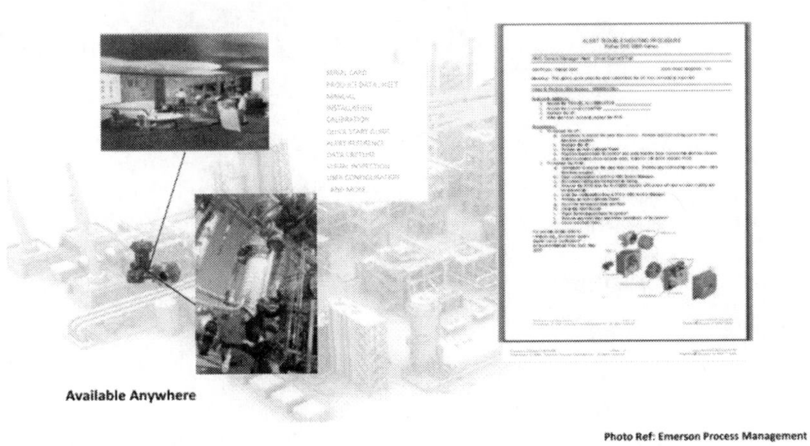

Move the data to the expert, not the expert to the data!

Plant Data Typical Challenges

- Data
  - What data should we collect?
  - How should we store it? Particularly non-numeric data?
  - How should we search and retrieve data of interest?
  - How should we get the best data in the right hands?
  - How should we then act on that data?
  - How do we do this quick and seamless with as little investment as possible?
- Platforms and Tools
  - What infrastructure should I use?
  - What tools should I use?
  - Should I store on-premise or in the cloud?
  - Should I use open source?
  - How do I scale?

What are the Components of a Predictive Analytics Implementation?

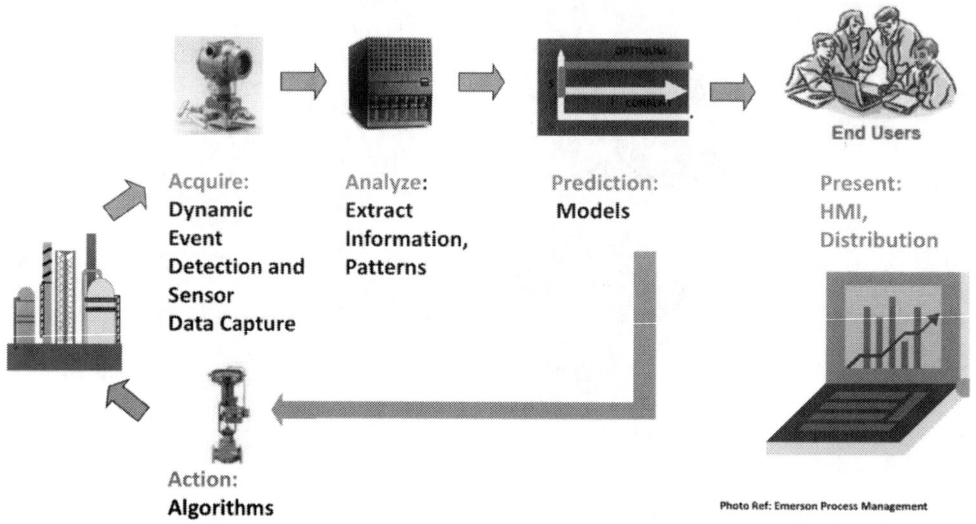

Move from reaction to events to anticipation and optimization

Major Predictive Analytics Techniques

- **Numerical Data (with adjustment for time series dependence)**
  - **Develop Predictive Model**
    - **Linear Models**
      - **Continuous Variables – *Generalized Linear Regression***
      - **Binary Outcome (Yes/No) – *Logistics Regression***
      - **Discrete Outcome (1,2,3,4,..) *Classification and Reference Trees (CART + Random Forest)***
    - **Non-Linear Models**
      - **Continuous Variables - *Neural Nets***
      - **Discrete Outcome (1,2,3,4,..) - *(CART + Random Forest)***
  - **Detect Anomaly**
    - ***Principal Component Analysis - PCA***
- **Textual Data**
  - ***Textual Analytics***
- **Spectral Data**
  - ***Nearest Neighbor***
  - ***Regression***
- **Video/ Images**
  - ***Neural Nets***

AFPM
American
Fuel & Petrochemical
Manufacturers

Making cents from data - where is the payback?

Plant Operating Objectives - The Four Zero's

- ***Safety*** **– the goal is <u>zero serious safety incidents</u>**
- ***Sustainability*** **– the goal is <u>zero significant environmental incidents, excess energy use and excess waste</u>**
- ***Reliability*** **– the goal is <u>zero unscheduled downtime</u>**
- ***Financial*** **- the goal is <u>zero lost profit opportunities</u>**

How can Predictive Analytics support these objectives?

Plant Decision Cycle

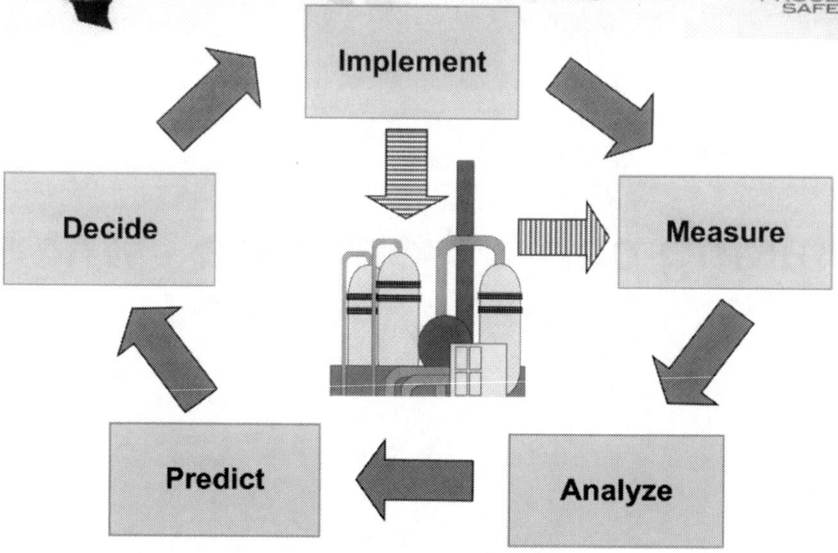

To have a financial impact predictive analytics has to improve this cycle – reduce delays, reduce uncertainty, etc.!

What impact can predictive analytics have?

- **Safety**
  - Avoiding incidents through early detection of potential hazardous situations
- **Availability/ Reliability**
  - Anomaly detection – identifying precursor events to unscheduled equipment outage or problems
  - Performance monitoring – detecting loss of process/ equipment performance before it impacts production capacity
- **Sustainability**
  - Comparing current usage of resources such as energy to its expected usage under current conditions and determining possible causes of variation
- **Financial Optimization**
  - Detecting and dissecting complex interacting constraints on production
  - Determining reasons for product quality/ yield issues
  - Understanding patterns and relationships – developing statistical models that explain them

AFPM
American
Fuel & Petrochemical
Manufacturers

Typical analytics project

Data from plants is often natively unsuitable for analysis

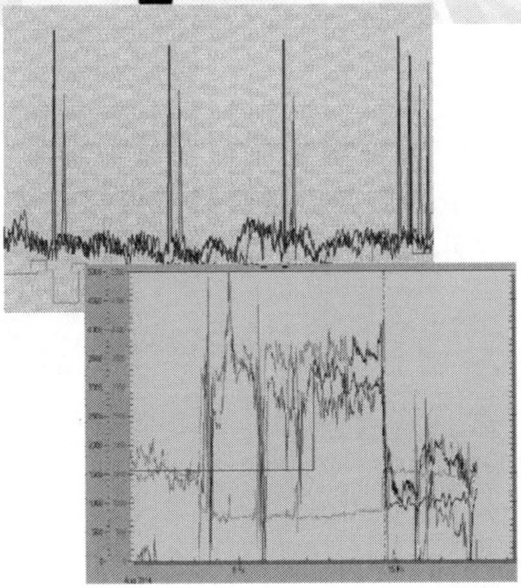

- We need to be able to preprocess and shape data before it can be used to generate models.
- These same techniques must be able to be used off-line to generate models and on-line to preprocess data before it is used to generate predictions, provide fault detections, and generate recommendations.

Cleaning Data – Outlier Detection

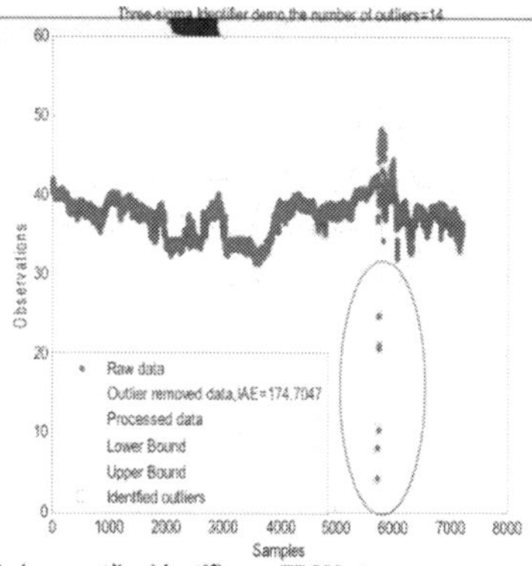

3-sigma outlier identifier on TD601 steam output data (moving window size=n/20=360)

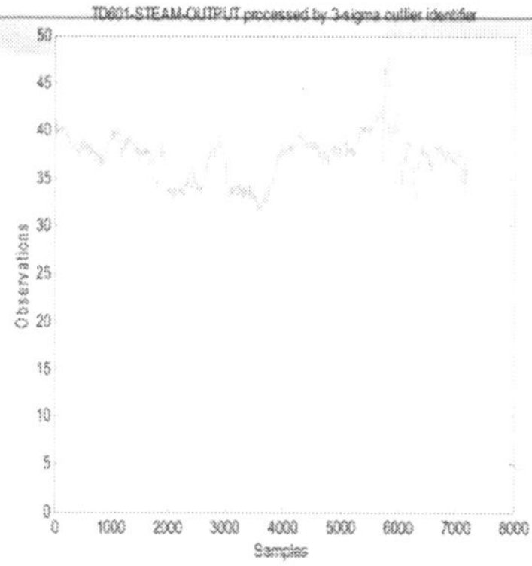

TD601 steam output data processed by 3-sigma outlier identifier(moving window size=n/20=360)

- Compressed data
- Missing data
- Outliers
- Noisy data

THE UNIVERSITY OF TEXAS AT AUSTIN
McKetta Department of Chemical Engineering
Cockrell School

Time Relationships Matter

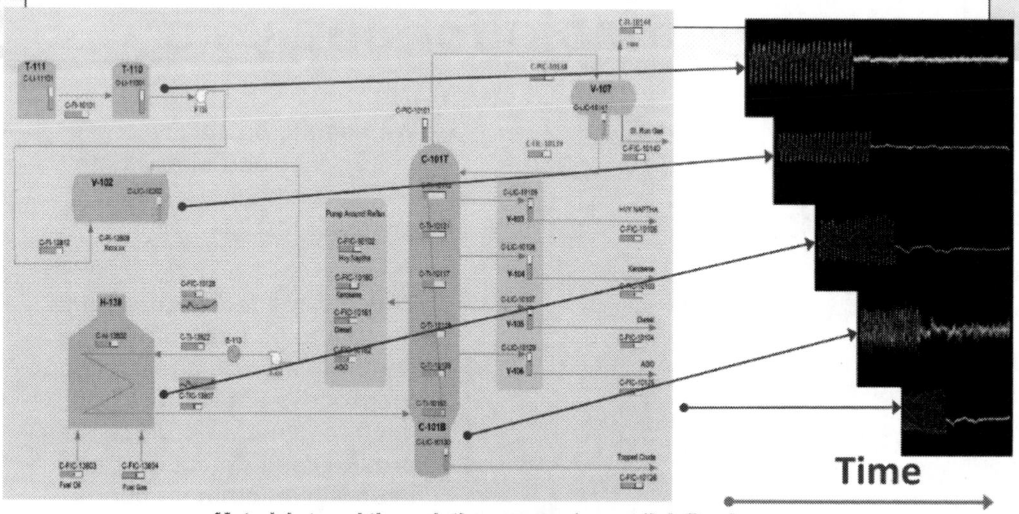

Materials travel through the process in a well defined arrangement. This flow of materials and the time delays are critical to being able to perform any kind of meaningful analysis.

By explicitly defining measurements and actuators along the flow path it is possible for an application to utilize this information to access and utilize information <u>in context</u>.

Finding sources of variations and oscillations

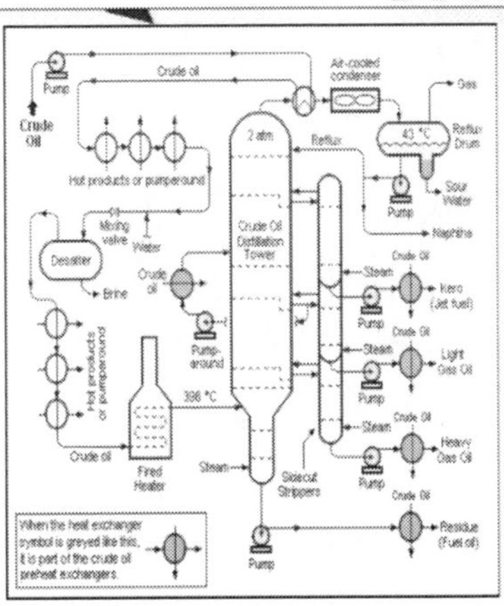

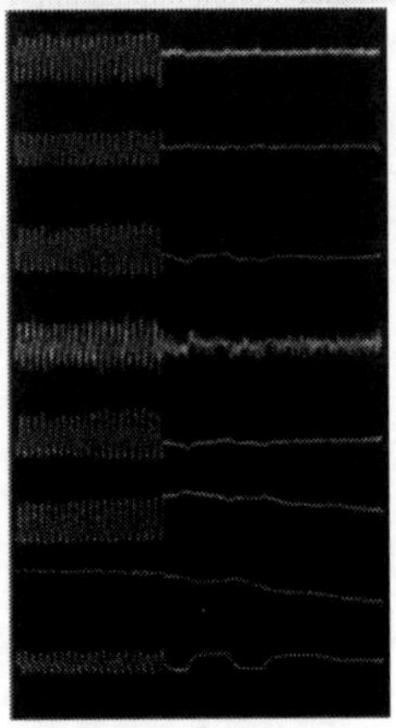

How do I know even where to look?

Parameters in processes are highly coupled. Variation in one parameter shows up in product quality and is reflected in many parameters downstream. Finding the cause can be difficult and time consuming.

Even knowing to look for it can often go undetected.

Cross Correlation and Causality

Output / Input

	X1	X2	X3	X4	X5
X1	0	2	3	2	4
X2	-2	0	1	0	2
X3	-3	-1	0	-1	1
X4	-2	0	1	0	1
X5	-4	-2	-1	-1	0

Table 1 Cross correlation time delay

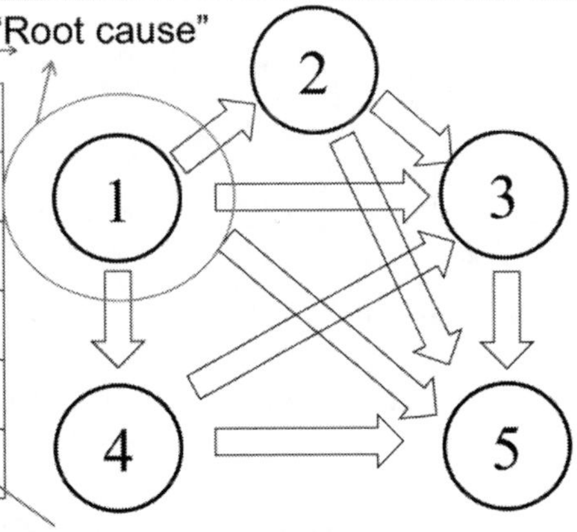

Figure 2 Process topology obtained by cross correlation

Correlation is not Causation!

Selecting parameters and time delays

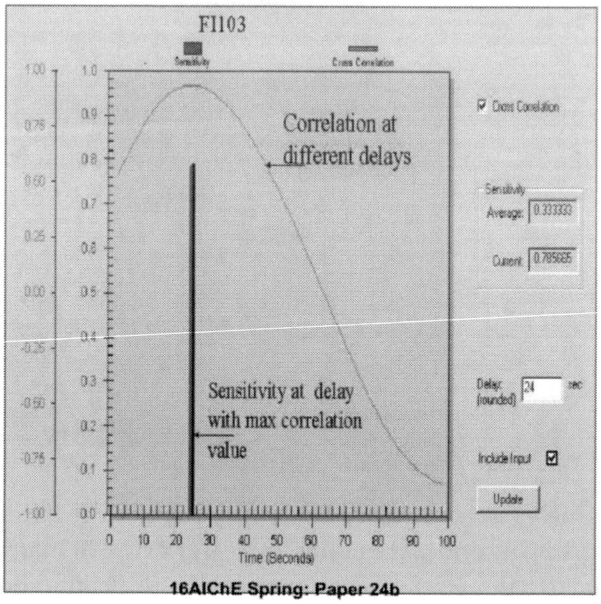

3/27/2016 16AIChE Spring: Paper 24b 27

Develop and Evaluate Models

Utilize Block Diagramming Environment to develop and evaluate models

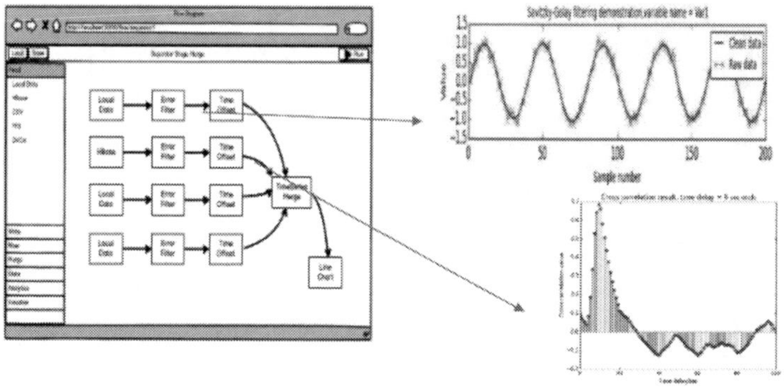

Interact with your model visualizing results at each stage of the data processing, when you are done, deploy the model for run-time execution

Visualizing Predictions

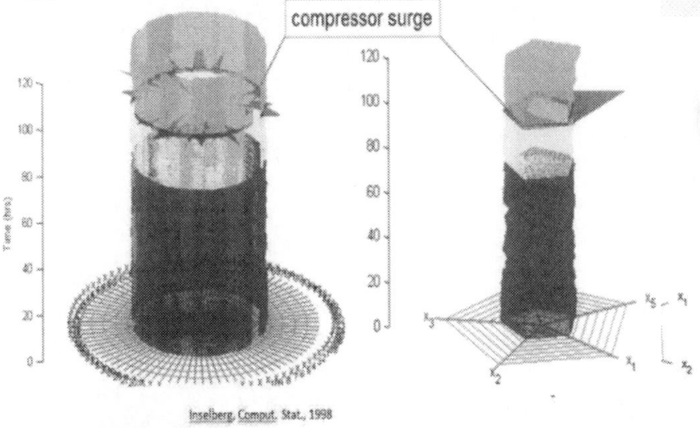

1 to 4 days before surge – Blue
1 day before surge – Yellow
2 hours before surge – Purple
Post surge - Green

Center for Operator Performance — An Industry/University Center for Excellence
The University of Texas at Austin — McKetta Department of Chemical Engineering — Cockrell School

Reference: Dunia, R; G. Rochelle; T.F. Edgar; M. Nixon; "Multivariate Modeling of a carbon dioxide removal process"; Computers & Chem Eng;60 (2014); pp. 381-395

AFPM American Fuel & Petrochemical Manufacturers

Case Studies

Case Study – Predictive Analytics – Waste Water Plant

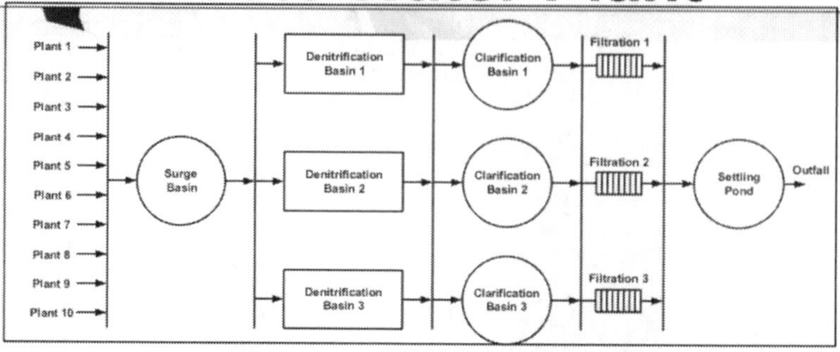

- Multiple effluent NH_3 spikes in outfall violating consent decree and incurring fines
- Residence time of the treatment plant >2 weeks making issue ID difficult
- More than 200 measured variables to be considered as possible predictors

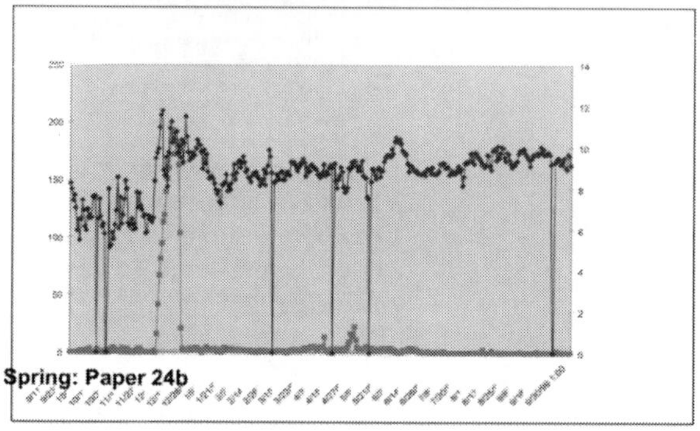

Case Study – Predictive Analytics – Waste Water Plant

- 5 years data analyzed for correlations with simultaneous identification of time lag
- Identification of strong correlation with pH of effluent from one of the plants – large drop in pH preceded excursion by one day
- Vessel cleaning correspondence with pH drop

Case Study – Predictive Analytics - Maintenance

- **Significance**
  - Sipchem Jubail site is considered a world-class complex with 4 large integrated plants
  - Successful implementation of complete integrated maintenance program
  - Built on installed base to create multi-phase asset management program
- **Project Scope**
  - Implementation of Asset Management including Reliability Centered Maintenance on 6,000 assets
  - Uses predictive analytics and diagnostics from field instrumentation
- **Audited Value**
  - 12% decrease in maintenance costs
  - 2% increase in Plant availability

Saudi International Petrochemical Company (Sipchem) actively develops and invests in petrochemical and chemical industries.

Sources: Abdullah Al Ghamde; John E. Hill,; *"Saudi International Petrochemical Company (Sipchem) Reliability Process,"* Paper 01A241; 2010 Emerson Global User Exchange; Sudheer Prabhakaran; John E. Hill,;*"Generating Actual Savings with Sipchem's Reliability Program,"* Paper 1B-2487; 2012 Emerson Global User Exchange

3/27/2016 16AIChE Spring: Paper 24b 33

Summary

- Predictive analytics is an evolving technology with many potential applications in the process industries
- Implementation of Predictive Analytics in the process industries has distinctive requirements/ issues due to the time series character and data
- Current applications in process fault detection, availability, safety and optimization have proven value – there will be more in the future

3/27/2016 16AIChE Spring: Paper 24b 34

Thank you for attending.
Questions?

3/27/2016 16AIChE Spring: Paper 24b 35

446613 ASTM D6299 and Leveraging Real-Time Analytics from the Lab

Abstract:

The EPA's new Tier 3 environmental standards require changes with employing ASTM D6299 to closely monitor emissions. This requires tighter control and deploying new analytic procedures. Frequently comparing control limits with T tests and F tests are adding steps which may be streamlined using automated procedures. We will also explore methods for automating lab instrument monitoring by streamlining the SQC duplicates and standards testing. Comparing real-time visibility of lab results with online analyzers can also provide powerful insights, but these projects face challenges with data alignment, and integrity. Issues with lead/lag relationships and data method variety such as sample vs event vs time based data, create complications for proper aggregation. These issues can be complicated with measurement errors, gaps in data, and noisy data. The session will showcase examples of how these challenges were overcome, and how additional value was achieved by being able to identify online instruments which may need attention. These methods provide the potential for refineries to run at higher levels of efficiency, reduce the need to downgrade product, and meet new regulatory standards.

444544 Analyzing the Process Image Based on Visual Saliency and Visual Clutter

Abstract:

As an important type of process data, images have been widely used in user-interface design and process monitoring because they can perform non-invasive analysis on the processes or products with low costs. However, an important issue with images is that they usually contain redundant information which can interfere with searching for the target item. For example, the process diagrams exhibited in the user-interface usually contain a large amount of equipment such as tanks and columns, and it is difficult for the user to find target items such as the reactor. A related topic is how to measure the negative influence of excessive items on the target searching, and based on the result we can improve the user-interface design by reducing the number and changing dyes of items. Another example is related to the flare analysis where the multivariate image analysis (MIA) has been established and widely used [1]. Isolating the flare flames in the images from the environment can be performed manually by visual detection. However, it is not quite efficient. Because the flare flames usually have distinct perceptual quality which makes them stand out from the environment, we can certainly apply advanced methods from computer science community to perform such a task automatically.

In this work, we apply methods including the visual saliency detection [2] and clutter analysis [3] from computer science community to analyze the process images to improve the user-interface design and facilitate the flare flames detection.

Reference

[1] D. ADDIN EN.REFLIST Castineira, B.C. Rawlings, and T. F. Edgar. Multivariate Image Analysis (MIA) for Industrial Flare Combustion Control. Industrial& Engineering Chemistry Research, 51(39): 12642-12652, 2012.

[2] J. Harel, C. Koch, and P. Perona. Graph-based visual saliency. In *Advances in neural information processing systems*, 545-552. 2006.

[3] R. Rosenholtz, Y. Li, L. Nakano. Measuring visual clutter. *Journal of Vision,7*(2):17, 2007.

444722 Extracting Relevant Quality Information through Image Data Analysis

Abstract:

As more powerful image analysis algorithms emerge from computer science and related fields, the application of digital image processing in manufacturing industries has been gaining popularity. There have been several reported case studies of image analysis for control and quality monitoring of industrial processes, such as detecting froth levels in flotation plants [1], industrial flare combustion control through multivariate image analysis [2], and spectral analysis using PCA techniques [3], [4]. However, most of these applications are proof of concepts in niche areas that monitor a single quality (i.e. position of the froth layer) using fast, efficient but relatively simple image analysis methods.

Inspired by these efforts, we addressed the challenge of measuring large amount of particle size distributions data through analyzing images of product samples. Using Hough transform based search algorithms [5], accurate quantitative information about the particle size distribution could be obtained. Scaling up the analysis algorithm to other product images allows us to create a particle evolution profile (particle size distribution over a set of experimental conditions) that could be used to discriminate against products of poor quality. This approach was applied to two industrial case studies.

In the first case, the product sample consists of spherical droplets in solution. Images of the droplets suspended in solution are taken while heating the solution to the desired temperature. The expansion and contraction profile of the droplets within the microscope image was hypothesized to be a key quality control parameter that needs to be investigated. In the second case, the quality of the crystalline structures is evaluated through image analysis as the process conditions of the reactor and centrifuge have been modified in order to reduce cycle batch time. The crystal length distribution was of interest and compared with length size distributions collected before these changes.

The image analysis based approach yielded successful results at a fraction of the cost of other conventional methods. Through analyzing the particle size distribution data, standard multivariate analysis techniques identified the most relevant information in the profile that could be used for future development of online in-situ monitoring techniques.

References

[1] S. Africa, P. O. Wits, and S. Africa, "Digital Image Processing As A Tool for On-Line Monitoring of Froth in Floatation Plants," *Miner. Eng.*, vol. 7, no. 9, pp. 1149–1164, 1994.

[2] D. Castin, B. C. Rawlings, and T. F. Edgar, "Multivariate Image Analysis (MIA) for Industrial Flare Combustion Control," 2012.

[3] L. L. Simon, K. Abbou, Z. K. Nagy, and K. Hungerbuhler, "Bulk video imaging based multivariate image analysis , process control chart and acoustic signal assisted nucleation detection," *Chem. Eng. Sci.*, vol. 65, no. 17, pp. 4983–4995, 2010.

[4] M. H. Bharati and J. F. Macgregor, "Multivariate Image Analysis for Real-Time Process Monitoring and Control," no. Figure 2, pp. 4715–4724, 1998.

[5] T. J. Atherton and D. J. Kerbyson, "Size invariant circle detection," vol. 17, no. February 1997, pp. 795–803, 1999.

444868 Use of Sparse Principal Component Analysis and Multidimensional Visualization Technique for Process Fault Detection and Diagnosis

Abstract:

Use of Sparse principal component analysis and multidimensional visualization technique for process fault detection and diagnosis

Shriram Gajjar, Murat Kulahci, Ahmet Palazoglu

University of California, Davis

Chemical process operations are typically subject to process or operational disturbances. Therefore, timely and effective detection and diagnosis of faults (monitoring) are critical to ensure safety and process stability and to maintain optimal levels of operation. For process monitoring, techniques based on first principle models have been studied for more than two decades but their contribution to industrial practice has not been pervasive due to substantial cost and time required to develop a sufficiently accurate model for a complex chemical plant. On the other hand, in a large-scale unit, a Distributed Control System (DCS) collects data from sensor arrays distributed throughout the plant and stores the data at high sampling rates. This data contains information about the underlying process characteristics and can be used for process monitoring. For process monitoring, a plant operator monitors several DCS screens and using his/her experience and domain knowledge, focuses on critical process variables to anticipate and prevent abnormal process operations. In the absence of such experience or domain knowledge, however, more automated techniques are required to inform and advise plant operators. A control chart is one of the primary techniques of statistical process monitoring of real time data. However, monitoring hundreds of variables simultaneously using univariate control charts is difficult. Moreover, 2-D charts limit our ability to visualize and interpret high-dimensional data. To overcome this challenge, Inselberg established the concept of parallel coordinates in 1985 [1]. In plane, parallel coordinates induce duality while in 2-D, they make cluster identification and pattern recognition easier. Moreover abundance of data and multiple variables to monitor at once make the task of monitoring unfeasibly difficult. Thus, principal component analysis (PCA), a technique that can capture the critical information in reduced dimensions, is widely used for process monitoring.

PCA-based monitoring methods, which build statistical models from normal operation data and partition the measurements into a principal component subspace (PCS) and a residual subspace (RS), are among the most widely used multivariate statistical methods [2]. In these approaches, the dimensions of the PCA model, i.e., the number of principal components retained, must be decided and this decision has an important role on the process monitoring performance. However, the approach to the determination of the number of PCs to be retained is not unique, especially due to the influence of sensor noise [3]. The choice of the number of PCs retained is a crucial step for the interpretation of monitoring results or subsequent analysis because it could lead to the loss of important information or the inclusion of undesirable interference. To tackle this challenge, a number of well-known techniques for selecting the number of PCs have been proposed. A simple approach is to choose the number of PCs for the explained variance to achieve a predetermined percentage, such as 85%, termed as cumulative percent variance (CPV) (Jackson, 1991). Other methods, including cross validation, average eigenvalue approach, variance of reconstruction error (VRE) criterion, and fault signal-to-noise ratio (fault SNR), have been proposed to determine the number of the retained PCs [3-6].

Using PCA for dimension reduction has drawbacks, that is, each PC is a linear combination of all m variables and the loadings are typically nonzero. Such nonzero loadings make it difficult to interpret the derived PCs. In order to get modified PCs with some possibly zero loadings, alternate approaches have been proposed by Cadima and Jolliffe [7], McCabe [8], Tibshirani [9] and Zou and Hastie [10]. In this paper we use the method proposed by Zou, Hastie [11] in which sparse loadings are obtained by imposing the lasso (elastic net) constraints on the coefficients (i.e. loadings) of the PCA model. In this method when the penalty (lasso) term vanishes, the results obtained are the exact PCA results. SPCA essentially is an optimization of the trade-off between variance captured by PCs and the sparsity. It allows the user to control the sparsity of the loadings and improve the ability to identify the important variables.

One of the challenges in using SPCA is in deciding the penalty parameters or choosing the number of non-zero variables/loadings. Zou, Hastie [11] use penalty parameters such that the sparse approximation explains almost the same amount of variance as the ordinary PCA does. In this paper, we vary the number of non-zero variables for each PC and until the variance captured by each PC in SPCA is approximately the same as the variance of the corresponding PC in the ordinary PCA. This method simplifies the process of selecting penalty parameters and provides a more intuitive solution for chemical processes. It also illustrates the key trade-off between sparsity and information retention.

Moreover, the fault detection ability has been shown to depend on the PCs retained in the PCA model [12]. Togkalidou, Braatz [13] also indicated that including components with smaller eigenvalues in the PCA model and excluding those with larger eigenvalues could improve the prediction quality. In most such approaches, not the magnitudes of component loadings but only significant data variations are considered to extract PCA components. The advantage of doing so is that this does not require making any a priori assumptions about data structure. The downside is, however, that the resulting loadings of the extracted PCA components are difficult to interpret. Motivated by this perspective, the present paper deals with several of the limitations inherently associated with the interpretation of loadings of retained PCs using PCA. In many cases, the traditional PCA can be altered in such a way that the obtained loadings would have a clear interpretation without significant loss of information extracted in each PC. Such an approach might help in the application of PCA as better understanding of the impact of PC loadings can clearly facilitate process monitoring.

In this paper, first we illustrate the advantages of using SPCA with a synthetic example. Second, we compare fault detection rates and diagnosis using SPCA and PCA for data obtained from the Tennessee Eastman benchmark process. In summary, this paper will focus on the use of parallel coordinates for multidimensional visualization using SPCA and discusses its accuracy for fault detection, fault diagnosis and fault propagation.

References

1. Inselberg, A., *The plane with parallel coordinates.* The Visual Computer, 1985. **1**(2): p. 69-91.

2. Cinar, A., A. Palazoglu, and F. Kayihan, *Multivariate Statistical Monitoring Techniques*, in *Chemical Process Performance Evaluation*. 2007, CRC Press. p. 37-71.

3. Tamura, M. and S. Tsujita, *A study on the selection of model dimensions and sensitivity of PCA-based fault detection.* Computers & Chemical Engineering, 2007. **31**(9): p. 1035-1046.

4. Valle, S., W. Li, and S.J. Qin, *Selection of the Number of Principal Components: The Variance of the Reconstruction Error Criterion with a Comparison to Other Methods.* Industrial & Engineering Chemistry Research, 1999. **38**(11): p. 4389-4401.

5. Wold, S., *Cross-Validatory Estimation of the Number of Components in Factor and Principal Components Models.* Technometrics, 1978. **20**(4): p. 397-405.

6. Dunia, R. and S. Joe Qin, *Joint diagnosis of process and sensor faults using principal component analysis.* Control Engineering Practice, 1998. **6**(4): p. 457-469.

7. Cadima, J. and I.T. Jolliffe, *Loading and correlations in the interpretation of principle compenents.* Journal of Applied Statistics, 1995. **22**(2): p. 203-214.

8. McCabe, G.P., *Principal Variables.* Technometrics, 1984. **26**(2): p. 137-144.

9. Tibshirani, R., *Regression Shrinkage and Selection via the Lasso.* Journal of the Royal Statistical Society. Series B (Methodological), 1996. **58**(1): p. 267-288.

10. Zou, H. and T. Hastie, *Regularization and variable selection via the elastic net.* Journal of the Royal Statistical Society: Series B (Statistical Methodology), 2005. **67**(2): p. 301-320.

11. Zou, H., T. Hastie, and R. Tibshirani, *Sparse Principal Component Analysis.* Journal of Computational and Graphical Statistics, 2006. **15**(2): p. 265-286.

12. Kano, M., et al., *Comparison of multivariate statistical process monitoring methods with applications to the Eastman challenge problem.* Computers & Chemical Engineering, 2002. **26**(2): p. 161-174.

13. Togkalidou, T., et al., *Experimental design and inferential modeling in pharmaceutical crystallization.* AIChE Journal, 2001. **47**(1): p. 160-168.

444315 Outlier Detection and Analysis in Batch and Continuous Processes

Abstract:

Title: Outlier Detection and Analysis in Batch and Continuous Processes

Author/Presenter: Peter J. Ryan, Ph.D., P.E.

Author/Presenter email: peter.ryan@responsepc.com

Company: Response Process Consulting LLC

Motivation

All manufacturing sectors, continuous or batch, gather process data for archiving and analysis purposes. Often the amount of data gathered, and the quality of data, makes it difficult to use this resource effectively. Major drawbacks in the quality of the available data include:

- large gaps in the process data
- noise, poor signal-to-noise ratios
- correlated data
- accuracy
- precision

New methods have been developed to handle these issues and to develop process models based on this resource. Specifically, a new approach to handling missing data and reconstructing quality data based on the observed (archived) data is presented. Once the models are developed, outliers can be identified in the continuous or batch data, and relationships between the Key Process Indicators (KPI's) and the upstream process variables can be established.

Approach

Often, steady state or dynamic models are available to describe a process. However, these first-principle models often do not have the granularity needed to describe product specifications such as color, turbidity or solvent loss due to entrainment in a separator. Machine learning can be used to examine large historical process data sets and determine the leading controlled variables of a process. The machines learning methods first fit the data to a specified model, and then use the model to explore the data space. Unsupervised learning gives the fitting method full ability to determine what is different in the data. Supervised learning requires the fitting method to consider both the archived process data and measured quality data (acquired off-line of the process). An example of unsupervised modeling is Principal Component Analysis (PCA). An example of supervised modeling is Partial Least Squares (PLS). Both of these methods can be used to model the historical process data and find outliers by plotting the two principal component elements (scores) that capture the most variability in the data. The score plots examine the data for clusters that classify the data as meeting product specifications and not meeting product specifications. The clusters that represent production runs not meeting product specifications can be further examined to discover the upstream process variables that are the cause of the product not meeting specification. While visual inspection has been described, statistical metrics such as the Squared Prediction Error (SPE) and Hotellings T^2 metrics can be calculated to find the same results in the data.

Results

Examples of modeling both continuous and batch processes are given. The continuous example is of a commodity chemical process where color is the product specification of interest. The batch example is of a nylon process where the relative viscosity is the product specification of interest. While the analysis methods are the same, one significant difference between handling continuous and batch data is that the batch data must first be "unfolded" before a model can be developed. Once the models are developed, clusters corresponding to successful and non-compliant products are found. The non-compliant product clusters are further examined, and the relationships between the product specification (KPI) and the upstream process variables causing the non-compliance are discovered. Figure 1 shows the results of the scores plot of the continuous commodity chemical example. Figure 2 is an example of the scores plot of the batch nylon example. In both cases, clusters of production activity where both in-spec and non-compliant quality production are observed. Focusing on the batch example, Figures 3 and 4 show the SPE and Hotelling T^2 charts of the initial process data. The outliers observed visually in Figure 2 are also detected numerically by calculating the Hotelling T^2 metric. The Hotelling T^2 metric finds points in the scores plot that are on the model plane but far away from the center-of-mass of the model. The SPE metric finds points in the clusters that are away from the model plane.

Contribution charts are used to discover the relationships between the KPI's and the upstream process variables, as shown in Figure 5. Note that in Figure 3 (SPE metric), batch 49 is far away from the model plane, even though its projection is in the cluster of points that represent batches with good product specification. The SPE contribution chart (Figure 5) reveals that deviations in five process variables caused the batch to product off-spec product. Specifically, the batch did not meet the turning points in the prescribed trajectories when the batch reached the 58th time interval of its run. An examination of the control system revealed that a control issue was accountable for missing the turning points.

These examples show how a very large set of process data – of varying quality (missing data, low signal-to-noise ratio, correlation, etc) – can be reduced in dimensionality and how the leading process variables can be identified and related to the downstream KPI's to improve product quality and consistency. The model has the granularity that is typically missing in first-principal models. The resource for this method of model building – historical process data – is readily available but seldom used. The resource is seldom used for process optimization and analysis because, without the methods needed to reduce the data-space dimensionality and cope with missing and correlated data, this resource presents too much raw, unconditioned information.

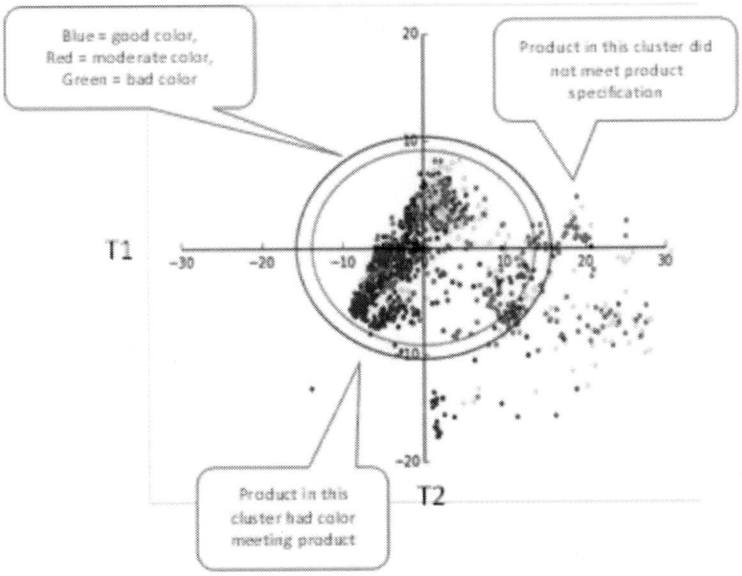

Figure 1: Model developed using **continuous** historical process data

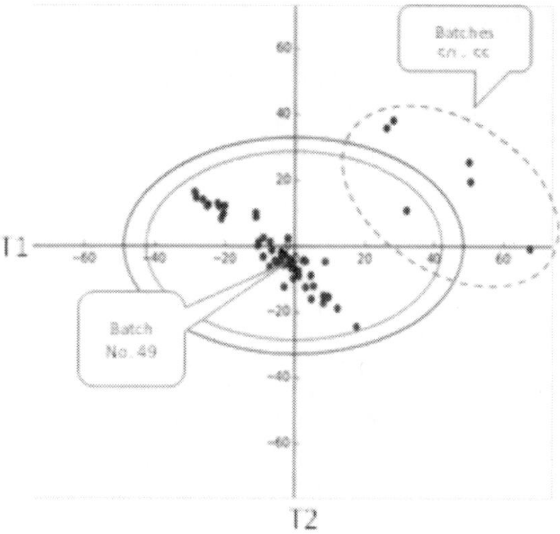

Figure 2: Model developed using **batch** historical process data

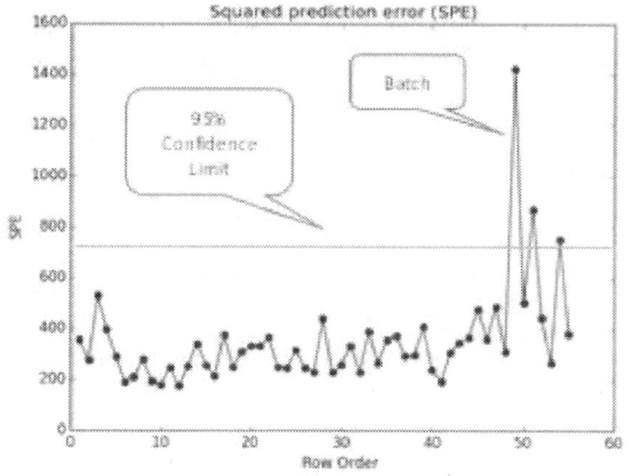

Figure 3: Squared Prediction Error (SPE) of the batch data (Nylon example)

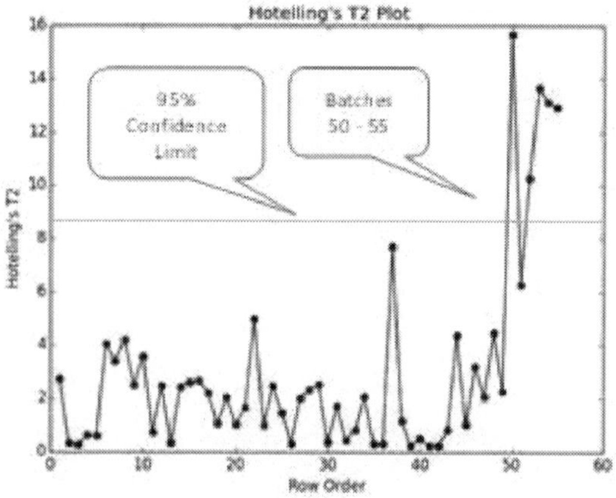

Figure 4: Hotellings T^2 metric of the batch data (Nylon example)

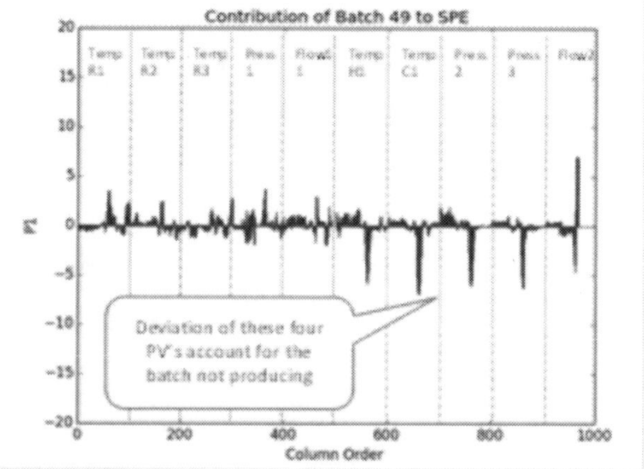

Figure 5: Contribution to the SPE of Batch 49 of the batch data (Nylon example)

(47b) Advances in Asset Health Monitoring and Predictive Analytics
By Tim Olsen from Emerson Process Management

Big data is a hot topic in many commercial and industrial applications, and it also applies to the operation of refineries and petrochemical plants through the use of sensor data to improve operations. Online data from sensors, often wireless, can be analyzed and used to provide plant operators with the actionable information to make timely decisions for improving asset reliability and efficient operations.

Big data from sensors can be analyzed by a variety of commercially available software programs, each of which can evaluate sensor data for abnormal operation and provide alerts. Properly trained operations are invaluable in this process, as they not only act on information from predictive analytics software, but often make decisions on their own based on sensor data.

Online data from sensors has been available for decades, but a transformation is now taking place due to the low cost and quick installation time for WirelessHART® (an international industry standard IEC 62591) sensors when compared to their wired counterparts. The term pervasive sensing describes the application of wireless sensors to areas which were formerly too expensive to monitor with wired instruments, leading to the creation of big data.

Examples of big data and predictive analytics applications include vibration sensing of rotating equipment, ultrasonic leak detection, steam trap monitoring, corrosion and erosion detectors, flows, pressures, and temperature measurements. These and other monitoring applications are providing major savings at many process plants in maintenance costs and energy while mitigating unscheduled downtime and potential ensuing safety and environmental incidents.

Here are some examples of monitoring applications utilizing asset health data and predictive analytics.

Vibration sensing
One of the earliest industrial applications of pervasive sensing was the use of small, battery-operated wireless vibration sensors to monitor the condition of motors, pumps, gearboxes and other non-critical rotating equipment. Readings from the sensors can be analyzed and alert the plant's asset management system when a piece of equipment shows abnormal operation or imminent failure.

A Midwest refinery needed to prevent hydrocarbon vapor cloud releases when pumps failed. Wireless vibration sensors now let operators know which pumps need service. They now monitor more than 150 pumps rather than manual checking non-critical pumps once a month. The system has provided early warning of three impending pump failures in its first year of operation.

Examining effects of corrosion
There are estimates that 36 percent of a typical refinery's maintenance budget is spent on corrosion remediation and repairs. Until recently, corrosion monitoring required a shutdown to install a monitoring device (e.g., detection coupons), and then subsequent shutdowns to inspect them. But the past few years have seen the introduction of wireless corrosion monitors that provide online data on corrosion rates that can be used to detect a spike in corrosion, control inhibitor injection, and correlate with other process data for root cause analysis. It is still recommended to continue using traditional coupons and probes along with inspections during plant shutdowns.

Monitoring heat exchangers
A Gulf Coast Refiner had been experiencing problems with a crude unit due to accelerated fouling within the preheat train. Installing wireless temperature and pressure instruments made it possible to use heat duty calculations to analyze heat exchanger rate of fouling and determining which bundle required cleaning; the return-on-investment was less than 90 days which didn't included the valuable knowledge learned about crude oil blend incompatibilities.

Monitoring of compressed air and gas leaks
The Department of Energy's Office of Energy Efficiency and Renewable Energy (EERE) states that leaks often waste as much as 20 to 30 percent of compressor output. EERE recommends the use of ultrasonic acoustic detectors to identify leaks; these can also detect leaks of process gases, including toxic and flammable gases, although in the latter two cases it is a good idea to back up the acoustic sensors with fixed gas detectors and optical flame detectors.

An excellent way to get a handle on waste in the plant is to monitor flow rates on all utility fluids: compressed air, steam, natural gas and water. Today this is easily done using wireless sensors, and the cost savings can be significant. In plants that already measure these flows, newer technology can reduce costs. Orifice plates in compressed air lines create permanent pressure losses; replacing them with averaging Pitot tube flowmeters saved a Brazilian chemical company $750,000 annually in electricity costs.

Conclusion
Advances in asset health monitoring and predictive analytics are being used today to mitigate unscheduled downtime and reduce maintenance costs by taking corrective action before an asset failure. Abnormal operation can quickly be identified and corrected with properly trained staff to act on this new timely analyzed information. Big data and predictive analytics will continue to have a role with shortages of skilled resources, thus having the staff be more efficient and effective at their role with additional information.

Using a Large Data Bank to Develop a Robust and Simple Model to Estimate the Saturation Pressure of Oil Reservoirs

By

Hassan Alkandari[1], Adel Elsharkawy[1], Adel Malallah, Osamah Alomair

Petroleum Engineering Department, College of Engineering & Petroleum
Kuwait University
Po Box 5969 Safat 13060-Kuwait

Abstract

Measurements of saturation pressure are crucial for all hydrocarbon reservoir fluids. Below the crude oil saturation pressure gas reaches a critical saturation; then, a two phase flow occurs and results in decreasing oil production and recovery. To maximize oil production and recovery, the reservoir pressure has to be maintained closer to the original saturation pressure. This pressure is normally measured using bottom-hole samples or surface recombined samples of oil and gas. Occasionally, the samples become unavailable and the pressure needs to be estimated using computational methods.

In this study, a large data bank of 231 crude oil compositions and saturation pressure measurements including literature data, unpublished data, and newly measured data were used to develop two empirical models to predict the saturation pressure of a variety of crude oils if the oil sample is unavailable or the experimental measurements are unreliable. The first proposed model utilizes the extended compositions of hydrocarbons up to the heptane plus fraction in addition to non-hydrocarbons. The second model uses the lumped compositions of light, intermediate, and heavy components in addition to non-hydrocarbon components as the input variables. The lumped model has the advantage of using fewer input parameters while maintaining the thermodynamics. The accuracy and validity of both models to calculate the saturation pressure for volatile oils, black oils, and heavy oils are presented using several compositional data. The models performance is also compared to the Peng-Robinson equation-of-state (PR-EOS), and the Soave-Redlich-Kwong equation-of-state (SRK-EOS) as well as all published methods that use compositions as input variables. The comparison indicates that the proposed models are much simpler and more accurate than the other computational methods. The proposed models treat the heptane-plus fraction as a single component, thus eliminating the splitting and characterizing the plus fractions and the binary interaction parameters needed for the EOS's calculations.

Key words: Saturation pressure (Ps), Bubble-point pressure (BPP), Equation of state (EOS), Correlations, and PVT Data

[1] Corresponding Authors: Hassan Alkandari , Hassan@hkandari.com, Adel Elsharkawy, adel.elsharkawy@ku.edu.kw

Introduction:

Reservoir fluid properties are an essential source of information for all calculations on hydrocarbon reserves. Accurate estimation of the pressure-volume-temperature (PVT) measurement is crucial. Of all the PVT properties, saturation pressure (Ps) is considered the most important parameter. It is well known that gas starts to flow once the reservoir pressure drops below the bubble point pressure and the gas saturation reaches a critical value. To minimize the gas flow and maximize the oil recovery, it is very important to maintain the reservoir pressure close to the original bubble point pressure.

Although the PVT properties are experimentally measured in laboratories and provide reliable results, this process is expensive, time consuming, and sometimes unavailable; therefore, measured field data (separator gas oil ratio, separator pressure, stock-tank oil specific gravity, and reservoir temperature) or reservoir fluid compositions are used to estimate the bubble point pressure.

There are two methods in the petroleum industry to calculate the saturation pressure of crude oil when experimental is unavailable; either using production data or compositional data.

Methods Based on Production Data (Composition is unavailable)

Numerous attempts have been made since the 1940s to develop correlations to estimate the saturation pressure from production data, such as producing gas oil ratio, oil gravity, gas gravity, reservoir pressure and temperature. Table 1 presents summary of all the methods that use production data to estimate the saturation pressure.

Methods based on Compositional data

Many authors have reported that the application of any correlation in other regions results in serious errors. These conclusions arise from the fact that regional crude oils represent a physical-chemical trend; therefore, all attempts to develop a general correlation using field measured parameters to estimate the saturation pressure from a large data bank have failed. For this reason, Elsharkawy (2003) made the first attempt to correlate the saturation pressure to the reservoir fluid composition instead of using the production data. In recent years, there have been several attempts to estimate the saturation pressure of oil reservoirs. These methods are reported in Table 1.

Elsharkawy (2003) presented a correlation of the bubble point pressure for the oil reservoir using the same input data needed for the equation-of-state (EOS) calculations, i.e., the compositional data and the reservoir temperature. Elsharkawy's model is based on the saturation pressure measurements of 60 crude oil samples from the Middle East and 75 collected from the literature. He demonstrates that the calculation of the saturation pressure depends on the type of EOS, the number of pseudo components, the method used to characterize the pseudo-components and the binary interaction number. Elsharkawy compared the performance of his model to the SRK-EOS and PR-EOS for black oils, volatile oils, heavy oils, and synthetic oils. He indicated that his model eliminates splitting the heavy plus fraction, characterizing the pseudo-components and inclusion of the binary interaction parameters needed for the EOS calculations.

Later, Tale et al. (2009) made the second attempt to model the saturation pressure of crude oil using 132 data samples published by Elsharkawy (2003). Tale's model, which is based on the alternating conditional expectation (ACE), has a set of 14 equations with 42 constants. The average absolute relative error (AARE) reported by Tale et al. indicates that the model did not perform well with the selected data. The ACE model has an AARE of 12.3% compared with AARE of 6.7% by Elsharkawy (2003) and 8.1% AARE for the SRK-EOS.

AlQuraishi (2009) attempted to model the saturation pressure using Elsharkawy's (2003) data. The model was developed using linear genetic programming. He correlated the saturation pressure to the formation temperature, methane content, and heptanes plus molecular weight. Thus, AlQuraishi's model neglected the effect of the intermediate and heptane plus content on the saturation pressure. The error analysis reported by AlQuraishi indicates that his model has a 5.8% error compared to 7.7% by Elsharkawy, 9.9% using the SRK-EOS, and 10.1% using the PR-EOS.

Bandyopadhyay and Sharma (2011) presented a model to predict the bubble point pressure, using temperature and fluid composition. They used 129 experimental and literature data originally published by Elsharkawy (2003) to develop their model. The model introduced a temperature interaction parameter that depends on compositions. The proposed model has a set of 24 coefficients and the estimated bubble point pressure is calculated as the sum of the series calculations. They reported an average absolute error (AAR) of 7.83% using their model compared to 8.3% using Elsharkawy's simple model (2003), 10.15% using SRK-EOS, and 10.58% using the PR-EOS.

Khamehchi et al. (2011) used an intelligent proxy simulator (IPS), which is based on an adaptive neural fuzzy inference system (ANFIS) and a structured artificial neural network (ANN) to predict the bubble point pressure and asphaltene onset pressure. The model uses a reservoir fluid composition and temperature to predict the Ps. It is not clear how the Ps is calculated from the proposed method, and they did not present the accuracy of the proposed method compared to the other published methods.

Farasat et al. (2013) presented a new mathematical approach to calculate the crude oil saturation pressure as a function of the temperature and the reservoir fluid composition. The model developed by Farasat et al used 130 experimental data, which were originally published by Elsharkawy (2003). Farasat used a support vector machine (SVM) to model the saturation pressure. They did not clearly demonstrate how to calculate the saturation pressure from the SVM model. Moreover, they did not indicate the overall average absolute deviation (AAD) for their model compared to other published models and both the PR-EOS and SRK-EOS for the 130 data.

Ahmadi et al. (2014) also presented an ANN technique called gene expression programming (GEP) to determine the bubble point pressure of the oil samples. Similar to Farasat et al. (2013), Ahmadi's model used most of the data previously published by Elsharkawy (2003) to construct a non-linear equation using reservoir fluid compositions and temperatures as input parameters. The proposed model is highly complicated and consists of 8 equations that have several constants compared to the one simple equation proposed by Elsharkawy (2003). They compared the performance of their proposed model to the SRK-EOS and PR-EOS; however, their paper does not mention how the heptanes plus fraction were treated and characterized for the EOS calculations. They reported an AAD of 4% using the proposed model compared to 8% using Elsharkawy's model (2003), 10.9% by PR-EOS, 10.5% by SRK-EOS, and 15% by Bandyopadhyay and Sharma (2011).

Gholami et al. (2014) estimated the bubble point pressure using a support vector regression SVR and a supervised learning algorithm SLT based on the statistical learning theory, and then the ACE was used. The model used reservoir fluid compositions and temperature to estimate the bubble point pressure. Gholami et al. used the same data originally published by Elsharkawy (2003). They presented three attempts based on genetic programming to correlate the bubble point pressure: SVR, ACE, and power-law committee machine (PLCM). The ACE-based method resulted in a set of three equations with a matrix of [8 by14] constants. Gholami et al. compared the accuracy of their model to the SRK-EOS and PR-EOS predictions reported by Elsharkawy (2003). Based on their AARD analysis, the PLCM model has the lowest errors followed by the SVR model, Elsharkawy's model, ACE model, PR-EOS, and SRK-EOS.

Lately, Jarrahian et al (2015) presented two empirical models to estimate the saturation pressure of black oils. The first model is based on production information. The second model is based on compositional information mostly from Elsharkawy (2003) and Heidaryan and Moghadasi (2011). They have not presented comparison of the accuracy of their model to Farasat (2013) model, Ahmadi (2014) model, and Gholami (2014) model.

This paper is concerned with the estimation of saturation pressure using compositional data. Therefore, the objective is to develop a robust yet simple model to estimate the saturation pressure for various types of reservoir oils using a larger data bank than that used by other published studies. The accuracy of the proposed models is compared to previously published models as well as SRK-EOS and PR-EOS.

Data bank

This study uses a large data set of 231 crude oil compositions and saturation pressure measurements. The data set consists of 130 data sets originally reported by Elsharkawy (2003); in addition to others collected from literature (Jacopy, 1958; Vogel, 1980; Coats, 1986; Droham, 1980; Dansh, 1992; Moharam, 1996; Rosenegger, 1999); 94 new data sets that have not been utilized before; as well as 7 newly measured samples at the Petroleum Fluid Research Center (PFRC) in Kuwait University. These data represents a wide range of compositions, such as volatile oils, black oils, and heavy oils. The data cover a pressure range of 537 to 5065 psia and a temperature range of 58 to 319 °F, crude oil compositions from C_1 to C_{7+}, non-hydrocarbons, such as N_2, H_2S, and CO_2, and the molecular weight and the specific gravity of the heptane plus fraction.

Data screening and classification

The compositions and the saturation pressure measurements in the data bank are subjected to a meticulous screening process to check the validity of the data. Our screening criteria uses the following steps: the sum of the molar compositions of all the components must add up to one, otherwise, the sample is eliminated. Samples with a missing molecular weight or specific gravity of the plus fraction are also eliminated.

After filtering, the data were reduced to 214 and divided into 179 measurements used to develop the new models, and 35 data sets are used for the blind test of the robustness of the proposed models. Statistical analysis of the input variables of the data set used in training and testing the extended and the lumped models are illustrated in Tables 2.

Several of the reported compositional data have extended analysis up to (C_{20+}). The following techniques are used for lumping or regrouping the extended analysis into single pseudo-fractions (C_{7+}). The mole fraction (X_i) of the individual components of the extended analysis is equal to the plus fraction (Ahmad, 2011).

$$X_{C_{7+}} = \sum_{i=7}^{n} X_i \qquad i=7, 8 \ldots n \qquad (1)$$

The average molecular weight (Mw) of the individual component of the extended analysis is equal to the molecular weight of the plus fraction.

$$Mw_{C_{7+}} = \sum_{i=7}^{n} \frac{X_i Mw_i}{X_{C_{7+}}} \qquad i=7, 8 \ldots n \qquad (2)$$

The density of the total plus-fraction can be calculated as follows:

$$\gamma_{C_{7+}} = \frac{X_{C_{7+}} Mw_{C_{7+}}}{\sum_{i=7}^{n} \frac{X_i Mw_i}{\gamma_i}} \qquad i=7, 8 \dots n \qquad (3)$$

It is important to note the significance of the density of plus fraction because its density determines its paraffinic naphthenic aromatic (PNA) content.

Equation-of-state calculations

The EOS calculation of the saturation pressure is compared with the newly developed models as well as the other methods. The two EOS calculations considered in this paper are the ones most considered and widely used in the petroleum industry to evaluate the volumetric and phase behavior of reservoir fluid properties: the Soave-Redlich-Kwong EOS (SRK-EOS) and the Peng-Robinson EOS (PR-EOS). It should be noted that the accuracy of any EOS calculation of the saturation pressure largely depends on the number of pseudo-components and the splitting schemes used to divide the heptane plus fraction; as well as the characterization and calculation of the critical properties of the pseudo-components and the binary interaction parameters. Therefore, the two EOS calculations of the saturation pressure are not the same, even though the number of pseudo-components, characterization of the pseudo-components, and binary interaction parameters were the same.

Model development

As previously stated, the objective of this work is to provide the petroleum reservoir engineer with a simple yet accurate tool to estimate the saturation pressure for a variety of crude oils using compositional data when experimental is unavailable. Two models are presented in this study to calculate the saturation pressure of the crude oil samples. The first model uses all input parameters, such as the molar compositions of the hydrocarbon and non-hydrocarbon components similar to the EOS models. This model is referred to as the extended model. The second model uses fewer input parameters in which some of the hydrocarbon components have been lumped into a single group; this is designated as the lumped model. Both models are simpler than all published correlations, physically correct, and have an acceptable accuracy compared with the well-known EOS.

Figure 1 shows a correlation matrix of all the data available under consideration. This figure also indicates the relative importance of each of the input variables into the first proposed model. It agrees with most published correlations and the EOS that the methane content, heptane plus content and temperatures have the greatest impact on the saturation pressure followed by the intermediate content. The data understudy indicates that the non-hydrocarbon has the smallest impact. This result is simply a reflection of the molar contribution of each component comprising the crude oil composition.

Model I - Extended Model

The extended model uses 13 variables comprising the molar compositions of the pure components, heptane plus properties, and reservoir temperatures, which are similar to the input parameters for the EOS. Figure 2 shows the correlation coefficient graph of all the input parameters in the extended model. Thus, the elimination of the intermediate effect has a negative impact on the accuracy of the predictive model. Our proposed extended model has the following form:

$$Ps = 124.72N_2 + 17.57CO_2 + 22.55H_2S + 64.22C_1 - 9.80C_2 - 52.49C_3 + 6.16C_4 - 19.22C_5 - 23.63C_6$$

$$- 21.43C_{7+} + 435.31\gamma_{C7+} + 1.14Mw_{C7+} + 4.29T \qquad (4)$$

Where Ps is the saturation pressure in psia, T is the temperature in °F, and the hydrocarbon and non-hydrocarbon compositions are expressed in mole.

Model II - Lumped Model

The second model is much simpler and uses only 7 -variables (the lumped model). Several lumping schemes are adapted in the petroleum industry for phase behavior modeling and reservoir simulation of the petroleum reservoir fluids (Ahmed, 2011). Hydrocarbon components are grouped into a pseudo fraction based on their molecular weight (Whitson, 1980), the similarity of their physical-chemical properties (Lee, 1979), their carbon number, their true boiling point or their number of families based on the chemical structure (Becker et al., 2015). (Emera and Samara, 2005) proposed lumping the hydrocarbon components into volatile (C_1 & N_2), intermediate (CO_2, H_2S, C_2 to C_4) and heavy categories (C_{5+}) to calculate the minimum miscibility pressure. (Duan et al., 2013) proposed lumping several components into a single carbon number (SCN) with predetermined values for properties, such as critical pressure, temperature, and eccentric factor. Others have proposed optimal lumping schemes based on the objectives (Lin et al., 2008). In lumped model, $C_{1+}N_2$, C_2, CO_2, and H_2S, the intermediate (C_3 to C_6), heptanes plus (C_{7+}) and its properties, and temperature are the input variable. The correlation matrixes of the lumped components are provided in Figure 3. Even though the non-hydrocarbon has the least effect their contribution is important, especially the H_2S in the case of sour crudes. Both models presented by AlQuraishi (2009) and Bandyopadhyay and Sharma (2011) have eliminated the effect of H_2S.

The proposed model has the following form:

$$Ps = 82.115*(C_1+N2)-11.635*C_2+39.158*CO_2+ 38.244*H_2S -1.217*(C_3—C_6) - 890.701*(C_{7+}/MW_{7+}*\gamma_{7+}) + 4.217 *T-1042 \quad (5)$$

Where Ps is the saturation pressure in psia; and T is the temperature in °F.

Result and discussion

Model accuracy

The accuracy of the two proposed models used to calculate the saturation pressures of the previously described crude oil samples is shown in Table 3. This table shows reports the average relative error, average absolute relative error standard deviation and the correlation coefficients for all the methods considered in this study. Table 3 also reports the accuracy of the PR-EOS, SRK-EOS, Elsharkawy's model, AlQuraishi's model, and Bandoypandyay and Sharma's model and Jarrahian model. The methods proposed by Gholami et al., Ahmadi et al., and Jarrahian et al are included in Table 3, even though they produce a unreasonable estimate of the saturation pressures, as explained in the appendix.

This table indicates that the extended and the lumped models have an (AARE) in the order of 8% and 8.5%, respectively. It should be noted that the complications in the Bandoypandyay and Sharma model did not improve the accuracy of the saturation pressure, because it had a large error. It is also noted that Jarrahian et al model produces an average error of 35%. Compared with the EOS, the lumped model has a few constants and eliminates the splitting, characterizing the pseudo-components and binary interaction parameters necessary for the EOS calculations. The lumped model as well as the extended model are very simple and require one-step calculations compared with all the other methods reported in Table 3.

A cross-plot of the calculated saturation pressure using the extended model based on the training data versus the measured data is illustrated in Figure 4A. This figure indicates that most of the data falls on the unit slope line, and there is no over-fit. This trend is supported by the high correlation of regression of the model (97.6%) presented in Table 3. Figure 4B shows a cross-plot of the testing data (28 from literature and 7 new measured in PFRC at Kuwait University). The extended model indicates an excellent agreement with all of the 28 data sets chosen at random, with the exception of a few data points at higher pressures. Additionally, the model has a perfect match for the 7 newly measured samples.

Figure 5A shows the performance of the lumped training-model for capturing the measured saturation pressure for most of the 179 samples. A cross-plot of the calculated saturation pressure using the lumped model based on the training data versus measured data is depicted in Figure 5B. This figure indicates that most of the data scatters falls to the unit slope line. The lumped model also has a high coefficient of regression of 97.4%, which is indicated in Table 3.

Models Validity

The accuracy and validity of the models for estimating the saturation pressure of various types of crude oil are reported in Table 4. This table illustrates the capability of the newly proposed models to predict the saturation pressure for volatile oils, black oils, and heavy oils. Table 4; provides the compositions, properties of the heptane plus fraction, temperature, and measured saturation pressures of the samples. Additionally, the calculated saturation pressure using the PR-EOS, SRK-EOS, Elsharkawy's model (2003), AlQuraishi's model (2009), and Bandoypandhyay and Sharma's model (2011), Jarrahian et al (2015); as well as the extended and lumped models are provided in Table 4. It is clear from this table that the proposed models have a wide range of applications regardless of crude oil types. We have also included information regarding the 7 newly measured data; at the PFRC, compositional, and saturation pressure. At the bottom of Table 4, we have provided the unpublished data of the compositions; and the saturation pressures at low temperatures. It is well known that the saturation pressure of a crude sample is measured at low temperatures in addition to the reservoir temperature as a method to check the quality of the sample.

Figure 6A reports the error of all the methods considered in this paper to predict the saturation pressure for volatile oils (heptane plus content below 30%). Both the extended and the lumped models have an AARE of nearly 4% followed by Bandoypandhyay and Sharma's model (2011); and Elsharkawy's model (2003); the SRK-EOS has nearly twice the AARE of the newly proposed model. AlQuraishi's model (2009) has an AARE of 12% because the model is based on methane and neglects the effect of heavy and intermediate components, whose contribution is highly significant in the reservoir and production performance of volatile crudes. Jarrhaine et al (2015) model is not suitable for volatile oils. It shows a 35% error in estimated saturation pressure.

The performance of the extended and lumped models compared to the various methods for the black oil samples (35< C_{7+} <55%) is provided in Figure 6B. The newly proposed models have an AARE close to 10%; followed by Bandoypandyay and Sharma's model 14%, the SRK-EOS and PR-EOS 16%, and Elsharkawy 17%. AlQuraishi's model has an AARE of 24%. Jarrhain et al (2015) model shows a 33% error in estimated saturation pressure even though the authors claim that the model is suitable for black oils.

Figure 6C shows the error level of all the methods as well as the proposed models for the newly measured crude oil samples at Kuwait University using the PFRC. Again, Jarrhain et al (2015) model show a 40% error in estimated saturation pressure.

The performance of all published models at low temperature is illustrated in Figure 6D. The low-temperature calculation indicates that Bandoypandhyay and Sharma's model (2011) has accuracy problems and is not valid at low temperatures because the authors proposed a temperature-composition related coefficient. Also, Jarrhain et al (2015) model is not suitable at low temperature. It shows a 40% error in estimated saturation pressure.

The ability of the proposed models to capture the thermodynamic is thoroughly checked against its independent variables. The validity of the model is compared to the experimentally measured data as well as the predictions by other methods used to capture the physics of increasing the saturation pressure as a function of temperature is illustrated in Figure 7. Two main points are observed here; first, the model matches the measured saturation pressures much better than the other methods. Second, it follows the physical trend predicted by the EOS's. Figure 8 shows the model performance for capturing the effect of changing the saturation pressure as a function of the molecular weight of the heptane plus fraction compared with the SRK-EOS and PR-EOS. The extended model follows the trend of the PR-EOS in increasing the saturation pressure. Elsharkawy's model (2003) was unable to predict the correct trend as predicted by the EOS's. The model sensitivity in capturing the change in the saturation pressure as a function of increasing the methane content is shown in Figure 9. The proposed model as well predicts the correct trend of increasing the saturation pressure when increasing the methane content.

Conclusion

In this paper, two models are presented to estimate the saturation pressure of a variety of crude oils when measures are unreliable or samples are unavailable. The two models are developed using compositional data and saturation pressure measurements from a large data bank comprising literature data, unpublished data, and newly measured and reported data of 231 crude oil samples. The input parameters in the first model are the extended hydrocarbon and non-hydrocarbon molar percentages and reservoir temperatures. The second model uses lumped hydrocarbon fractions of intermediate, and heptane plus fractions in addition to its properties; and the reservoir temperature as input parameters. The accuracy of the two models has been tested and compared to that of the SRK-EOS and the PR-EOS; as well as published correlations that use compositional data as the input parameters. The comparison indicates that the two models are much more accurate than all the published methods. The validity of both models has also been tested using various types of crude oils samples and compared to experimentally measured saturation pressures and estimates using the EOS as well as the published correlations. The proposed models are physically correct and much simpler than all the published methods. The models do not require tedious computations and eliminate splitting of the plus fraction, characterization of the pseudo-components; as well as the binary interaction parameters that are required for the EOS calculations.

Acknowledgement

The authors thank the General Facility Lab at Kuwait University for their cooperation and conducting the compositional and saturation pressure measurements of the 7 new samples described in this paper using Petroleum Fluid Research Center (PFRC) project no. GE 01107.

References

Agarwal, R.K., Li, Y.K., Nghiem, L., 1990. A regression technique with dynamic parameter selection for phase – behavior matching. SPE Res. Eng. February, 115 – 120.

Ahmadi, M. A., Zendehboudi, S, James, L., Elkamel, A., Dusseault, M., Chatsiz, I., Lohi, A., 2014, New Tools to determine Bubble Point Pressure of Crude Oils: Experimental and Modeling Study, *Journal of Petroleum Science and Engineering, v.123, 204-216.*

Ahmed, T., 2011, Splitting and lumping schemes of the plus fraction, working guide to vapour-liquid phase equilibria calculations, Chapter 9, Hydrocarbon Phase Behaviour, Gulf Publishing Co. 111-126.

Al-Gathe, A., Abd-el Fattah, Kh., El-Mtwally, K., 2014, New artificial intelligent approach for bubble point pressure, Int. J innovation and Sci. Research, Vol.4, No.2, July, 121-135.

Al-Marhoun, M.A. 1985. Pressure-Volume-Temperature Correlations for Saudi Crude Oils. Presented at the SPE Middle East Oil Technical Conference and Exhibition, Bahrain, 11–14 March. SPE-13718.

Almehaideb, R.A. 1992. Improved PVT Correlations for UAE Crude Oils. Presented at the Middle East Oil Show and Conference, Bahrain, 15-18 March. SPE-37691.

Al-Najjar, H.S., Al-Soof, N.B.A., and Al-Khalisy, K.M. 1988. Correlations for Bubble-Point Pressures, Gas Oil Ratios and Formation Volume Factors for Iraqi Crude Oils. *Journal of Petroleum Research* (June 1988): 13.

AlQuraishi, A. A., 2009, Determination of crude oil saturation pressure using linear genetic programing, Energy & Fuel, 23, 884-887.

Al-Shammasi, A.A. 2001. A Review of Bubblepoint Pressure and Oil Formation Volume Factor Correlations. *SPE Res. Eval. & Eng.* 4 (2): 146-160. SPE71302.

Asgarpour, S., McLauchlin, L., Wong, D. et al. 1989. Pressure-Volume-Temperature Correlations for Western Canadian Gases and Oils. *J Can Pet Tech.* 28 (4): 103.

Asoodeh, M., Bagheripour, P.,2012, Estimation of bubble point pressure from *PVT* data using a power-law committee with intelligent systems, *J. Pet. Sci. and Eng., Vol. 90–91, July 2012, Pages 1-11.*

Bandyopadhyay, P., Sharma, A., 2011, Development of a new semi analytical model for prediction of bubble point pressure of crude oils, *Journal of Petroleum Science and Engineering, Volume 78, Issues 3–4, September 2011, Pages 719-731.*

Becker, P. J., Celse, B., Guillaume, D., Dulot, H., Costa, V., 2015, Hydrotreatment modelling for variety of VGD feedstocks: a continuous lumping approach, Fuel, 139, 133-143.

Coats, K.H., and Smart, G.T., 1986, Application of regression base EOS PVT Program to laboratory data, SPERE, May 277-299.

Dansh, A. Xu, D. H., Todd, A. C., 1992 A grouping method to optimize reservoir description for compositional simulation of gas injection process, SPERE, Aug. 343-348.

De Ghetto, G. and Villa, M. 1994. Reliability Analysis on PVT Correlations. SPE-28904, Presented at the European Petroleum Conference, London, United Kingdom, 25-27 October.

Dindoruk, B. and Christman, P.G. 2001. PVT Properties and Viscosity Correlations for Gulf of Mexico Oils. Presented at the SPE Annual Technical Conference and Exhibition, New Orleans, 30 September-3 October. SPE-71633

Dokla, M.E. and Osman, M.E. 1992. Correlation of PVT Properties for UAE (United Arab Emirates) Crudes. *SPE Form Eval.* (March 1992): 41.

Droham, J.K., 1988, Enhancing the evaluation of PVT data, OSEA paper presented the 7th Offshore South East Asia, Feb. 2-5, Singapore.

Duan, J., Wang, W., Liu, H., Cong, J., 2013, modelling the characterization of the plus fraction by using continuous distribution function, Fluid Phase Equilibria, 345, 1-10.

Elam, F.M. 1957. *Prediction of Bubble Point Pressures and Formation Volume Factors from Field Data.* MS thesis, University of Texas at Austin, Austin, Texas.

Elmabrouk, S., Zekri, A., Shirif E., 2014, The prediction of bubble point pressure and bubble point oil formation volume factor in absence of PVT analysis, PET. Sci. & Tech., 32, 1168-1174.

Elsharkawy, A. M., 2003, An empirical model for estimating the saturation pressures of crude oils, Journal of Petroleum Science and Engineering 38 (2003) 55-77.

Elsharkawy, A.M. and Alikhan, A.A. 1997. Correlations for predicting solution gas/oil ratio, oil formation volume factor, and undersaturated oil compressibility. *J. Pet. Sci. Eng.* 17 (3–4): 291-302.

Emera, M., Samara, H., 2005, Use of genetic algorithm to estimate CO2-oil minimum miscibility pressure- a key parameter in design of CO2 miscible flood, J. Pet. Sci. Eng., Vol. 46, issue 1-2, 37-52.

Farasat, A., Shokrollahi, A., Arabloo, M., Gharagheizi, F., Mohammadi, A., 2013, toward an intelligent approach for determination of saturation pressure of crude oil, Fuel Processing Technology 115 (2013) 201-214.

Frashad, F., LeBlanc, J.L., Garber, J.D. et al. 1996. Empirical PVT Correlations for Colombian Crude Oils. SPE-36105 Presented at the SPE Latin American and Caribbean Petroleum Engineering Conference, Port of Spain, Trinidad and Tobago, 23–26 April. -MS.

Gharbi, R.B.C., Elsharkawy, A.M., 1999, neural network model for estimating the PVT properties of Meddle East crude oils, SPE Reservoir Eval. Eng. 2 (3), 255-265.

Gholami, A., Asoodeh, M., Bagheripour, P., 2014 How committee machine with SVR and ACE estimates bubble point pressure of crudes, *Fluid Phase Equilibria, Volume 382, 25 November 2014, Pages 139-149.*

Glasø, Ø. 1980, Generalized Pressure-Volume-Temperature Correlations. *J Pet Tech.* 32 (5): 785-795.
Heidaryan, E., Moghadasi, J., 2011, Crude saturation pressure through imperial correlation, Thermodynamics 2011, Athens, 31 August- 3 September 2011.
Hasan, M., 1993, PVT Correlations for Indonesian Crude Oils. *Proc.* 22nd Annual Convention of the Indonesian Petroleum Association, Indonesia, Vol.1, IPA93-2.1-070.
Jarrhian, A., Moghadsi, J., Heidaryan, E., 2015, Empirical estimating of black oil bubblepoint (saturation) pressure, J. Petroleum Science and engineering, (126), 69-77.
Jacopy R.H., Berry V.J.,1958, A method for predicting the pressure maintenance for reservoir producing volatile oil, AIME 213, 59-65.
Kartoatmodjo, T., Schmidt, Z. 1994, Large Data Bank Improves Crude Physical Property Correlations. *Oil Gas J.* 92 (27): 51–55.
Kartoatmodjo, T.R.S. and Schmidt, Z. 1991, *New Correlations for Crude Oil Physical Properties*, SPE 23556.
Khairy, M., El-Tayeb, S., Hamdallah, M., 1998. PVT Correlations Developed for Egyptian Crudes. *Oil Gas J.* 96 (18): 114.
Khamekhcki, E., Behvandi, R., Rashid, F., 2011, Prediction of bubble point pressure & asphaltene onset pressure during CO2 injection using ANN & ANFIS models, Pet. Sci. & Tech. Vol. 1 No.2, 35-45.
Labedi, R.M. 1982, *PVT Correlations of the African Crudes*, PhD thesis, CSM.
Labedi, R.M. 1990, Use of Production Data to Estimate the Saturation Pressure, Solution GOR, and Chemical Composition of Reservoir Fluids, SPE-21164-MS. Presented at the SPE Latin America Petroleum Engineering Conference, Rio de Janeiro, Brazil, 14-19 October,
Lasater, J.A. 1958, Bubble Point Pressure Correlations. *J Pet. Tech.*10 (5): 65–67. SPE-957-G.
Levitan, L.L. and Murtha, M. 1999. New Correlations Estimate Pb, FVF. *Oil Gas J.* 97 (10): 70.
Lee, S., 1979, Experimental and theoretical studies on the fluid properties required for simulation of thermal process, Paper SPE 8393.
Li, Y.K., Nghiem, L.X., Siu, A., 1985. Phase behavior computations for reservoir fluids: effect of pseudo-components on the phase diagrams and simulation results. J. Can. Pet. Technol. Nov. – Dec., 29 – 36
Lin, B., Leibovici, C. F., Jorgensen, S. B., 2008, optimal component lumping: problem formulation and solution techniques, Computer & Chem. Eng., 32, 1167-2272.
Macary, S.M. and El-Batanoney, M.H. 1992, Derivation of PVT Correlations for the Gulf of Suez Crude Oils. *Proc. of the* 11th EGPC Pet. Exp. & Prod. Conference, Vol. 1, 374, Cairo, Egypt,
Moharam, H.M., 1995. Prediction of viscosity of heavy petroleum fractions and crude oil using a corresponding states method. Ind. Eng. Chem. Res. 34, 4140 – 4144.
Moradi, B., Malekzadeh, E., Amani, M., Boukadi, F.H., Kharat, R., 2101, Bubblepoint pressure imperial correlation, SPE 132756.
Obomanu, D.A. and Okpobiri, G.A. 1987, Correlating the PVT Properties of Nigerian Crudes. *J. Energy Res. Tech.* 109 (4): 214-217.
Okoduwa, I. G., Ikiensikimama, S.S., 2010, Bubble point pressure correlation for Niger Delta Crude Oils, SPE 136968.
Omar, M.I. and Todd, A.C. 1993. Development of New Modified Black Oil Correlations for Malaysian Crudes. SPE-25338 Presented at the SPE Asia Pacific Oil and Gas Conference, Singapore, 8-10 February.
Owolabi, O.O. 1984. *Reservoir Fluid Properties of Alaskan Crudes*. MS thesis, University of Alaska, Fairbanks, Alaska (May 1984).

Pedersen, K.S., Thomassen, P., Fredenslund, Aa., 1988a. Characterization of gas condensate mixtures. AIChE Spring National Meeting, March 6 – 10, New Orleans, LA.

Peng, D. y., Robinson. D. B., 1976. A new two-constant equation of state , Ind. Eng. Chem. Fund. 15, 59-63.

Petrosky, G.E. Jr. 1990. *PVT Correlations for Gulf of Mexico Crude Oils*. MS thesis, University of Southwestern Louisiana, Lafayette, Louisiana.

Petrosky, G.E. Jr. and Farshad, F. 1998. Pressure-Volume-Temperature Correlations for Gulf of Mexico Crude Oils. *SPE Res Eval & Eng.*1 (5): 416-420. SPE-51395.

Riemens, W.G., Schutle, A.M., Jong, L.N.J., 1988. Birba field PVT variations along the hydrocarbon column and confirmatory field tests. J. Pet. Technol. Jan., 83 – 88

Rosengger, L., Wu., R.1990, Integrated PVT data characterization- lessons from four case histories, J. Can. Per. Tech., 38, (13), (paper 97-05)

Shojaei, M., Bahrami, F., Barati, P., Riahi, S., 2014, Adaptive neuro-fuzzy approach for reservoir oil bubble point pressure estimation, *J. of Natural Gas Sci. and Eng., Volume 20, September 2014, Pages 214-220.*

Soave, G. 1972. Equilibrium constants from a modified Redlich-Kwong equation of state, Chem. Eng. Sci. 27, 1197-1203.

Standing, M.B. 1947. A Pressure-Volume-Temperature Correlation for Mixtures of California Oils and Gases. *API Drilling and Production Practice* (1947): 275-287.

Tale, M., Alquraishi, A., Al Mushaigeh, E. 2009, New model for crude oil saturation pressure prediction using alternation conditional expectation, Offshore Mediterranean Conf. & Exhibit. Italy 25-27, 2009.

Talebi, R., Ghiasi, M. Talebi, H., Mohammadyian, M., Zendehboudi,S., Arabloo, M., Bahadori, A., 2014, Application of soft computing approaches for modelling saturation pressure of reservoir oil, *J. Nat. Gas Sci. and Eng., Vol. 20, Sep. 2014, 8-15.*

Vazquez, M., Beggs, H. D. 1980, Correlations for Fluid Physical Property Prediction. *J Pet. Tech.* 32 (6): 968-970. Vazquez, M.E. 1976. *Correlations for Fluid Physical Property Prediction,* MS thesis, University of Tulsa.

Velarde, J., Blasingame, T.A., and McCain Jr., W.D. 1997. Correlation of Black Oil Properties at Pressures below Bubble Point Pressure - A New Approach. Presented at the Annual Technical Meeting of CIM, Calgary, Alberta, 8–11 June. PETSOC 97-93.

Vogel, J.R., Yarborough, L. 1980, the effect of nitrogen on phase behavior and phase properties of reservoir fluids, SPE 8815.

Whitson, C. H., 1980, Characterizing hydrocarbon plus fraction, SPEJ, 24 (4), 683-694.

Williams, C.A., Zana, E.N., 1980. Use of the Peng-Robinson equation of state to predict hydrocarbon phase behavior and miscibility for fluid displacement. SPE 8817, SPE/DOE on Enhanced Oil Recovery, April 20–23, Tulsa, OK.

Appendix

In this appendix, sample calculations are presented to show the deficiency of some of the published methods.

Input parameters:

N_2	CO_2	H_2S	C_1	C_2	C_3	C_4	C_5	C_6	C_{7+}	GR.C7+	MWC7+	T, °F	Ps, psia
0.65	0.02	0	45.02	12.45	8.93	6.03	3.02	1.44	22.44	0.81	184	140	3002

Ahmadi et al (2014)

Ahmadi et al. proposed an 8 steps calculation (copied from their paper) to estimate the BPP as follows:

$$BPP = A \times Mw_{C7+} - \frac{B}{C_{7+}} + C \times Vol./Inter. + D \times SG_{C7+} + \frac{E}{SG_{C7+}} + \frac{F}{Mw_{C7+}} + T \times 505.967 + \frac{2.45789 \times 10^6}{T} + \frac{G}{Vol./Inter.} \quad (5)$$

$$A = 1488.46 - SG_{C7+} \times 1087.64 - \frac{601.512}{SG_{C7+}} + 1670.58 + T \times 0.40771 + \frac{6345.73}{T} + \frac{8.72863}{Vol./Inter.} \quad (6)$$

$$B = \frac{4132.53}{Vol./Inter.} + \frac{6.1977 \times 10^7}{Mw_{C7+}} + T \times 3181.39 + \frac{1.06393 \times 10^8}{T} \quad (7)$$

$$C = \frac{20964.7}{C_{7+}} - \frac{1.76162 \times 10^6}{Mw_{C7+}} - T \times 19.1963 - Mw_{C7+} \times 21.5427 + \frac{12851.5}{SG_{C7+}} - \frac{712613}{T} - 3477.49 + SG_{C7+} \times 9532.73 \quad (8)$$

$$D = 1.05345 \times 10^{+06} - \frac{1.50748 \times 10^8}{Mw_{C7+}} - T \times 676.048 - \frac{1.20252 \times 10^7}{T} - \frac{12178.2}{Vol./Inter.} \quad (9)$$

$$E = 641357 - \frac{1.02891 \times 10^8}{Mw_{C7+}} - T \times 301.034 - \frac{1.91446 \times 10^6}{T} - \frac{3873.62}{Vol./Inter.} \quad (10)$$

$$F = 2.51471 \times 10^{+08} + T \times 25332.5 + \frac{3.47893 \times 10^8}{T} - \frac{244573}{Vol./Inter.} \quad (11)$$

$$G = 20125.6 - T \times 23.9271 - \frac{703193}{T} \quad (12)$$

Ahmadi et al. define the Vol. /Inter. as the mole percentage ratio of the volatile (C_1& N_2) to the intermediate components (C_2 to C_6, CO_2 and H_2S). Application of the above equations from their paper using the input parameters results in the following:

Vol./Inter.	A	B	C	D	E	F	G	BPP
1.43211	1643.9	1545062	-271.127	45122.71	23642.47	2.57E+08	11753	1,794,123

The calculated saturation pressure (BBP) from Eq.5 is 1,794,123 psia which is unreasonable. This indicates a deficiency in the model.

Gholami et al (2014)

Gholami et al. presented 3 steps calculation (copied from their paper) to estimate the BPP as follows:

$$Tr(M) = \sum_{i=0}^{3} B_i M^i \quad (15)$$

Application of Eq. 15 resulted in the following

N_2	0.00584
CO_2	0.060917
H_2S	0.050636
C_1	-0.47658
C_2	-0.41547
C_3	-0.45083
C_4	-0.02003
C_5	0.068769
C_6	0.273921
C_{7+}	1.814523
SG.C7+	-0.0197
MWC7+	37.03525
T., F	-0.11139

Application of Eq. 16

$$Tr(P_b) = Tr(N_2) + Tr(CO_2) + Tr(H_2S) + Tr(C_1) + Tr(C_2) + Tr(C_3) + Tr(C_4) + Tr(C_5) + Tr(C_6) + Tr(C_{7+}) + Tr(SG_{C7+}) + Tr(MW_{C7+}) + Tr(T) \quad (16)$$

Resulted that Tr(Pb) = 37.8159

$$P_b = (57.0840387741 \times (Tr(P_b))^2) + (1166.9102304199 \times (Tr(P_b))) + 2207.3573204358 \quad (17)$$

Application of Eq. 17 resulted in saturation pressure (P_b) = 127,967 psia, which is unreasonable, compared to the measured one is 3002 psia. This indicates that there is a deficiency in the model.

Jarrahian et al (2015)

Jarrahian et al. proposed the following Eq. (copied from their paper) to estimate the saturation pressure:

$$P_{sat} = A_0 \exp\left(A_1 z_{N_2} + A_2 z_{CO_2} + A_3 z_{H_2S} + A_4 z_{C_1}^{A_5} + A_6 \sum_{2}^{6} z_{C_i} + + A_7 z_{C_{7+}}^{A_8} \gamma_{C_{7+}}^{A_9} Mw_{C_{7+}}^{A_{10}} + A_{11}(T - 459.67)\right) \quad (2)$$

Application of Eq. 2 resulted in saturation pressure (Ps) = 1812 psia, compared to the measured one is 3002 psia. This indicates that model has inaccuracy problem. This has been the pattern in all the crude oil samples shown in Table 4 and figures 6A through 6D.

Table 1- Summary of attempts to calculate the saturation pressure of crude oils

Method based on production information	Method based on compositional data
Standing, 1947;	Elsharkawy , 2003
Elam, 1957;	Tale et al., 2009
Lasater, 1958;	AlQuraishi, 2009
Vasquez, 1976;	Bandyopadhyay and Sharma, 2011
Vasquez and Beggs, 1980;	Khamehchi et al., 2011
GlasØ, 1980;	Farasat et al., 2013
Labedi, 1982&1992;	Ahmadi et al., 2014
Owolabi, 1984;	Gholami et al. 2014
Al-Marhoun, 1985;	Jarrahian et al 2015
Obomanu and Okpobiri, 1987;	
Asgarpour et al., 1989;	
Al-Najar et al., 1988;	
Dokla and Osman, 1990;	
Petrosky, 1990;	
Kartoatmodjo and Schmidt, 1991&1994;	
Macary and El-Batanoney, 1992;	
Omar and Todd, 1993;	
Hasan, 1993;	
De Ghetto and Villa, 1994;	
Farashad et al., 1996;	
Almehaidaeb, 1997;	
Elsharkawy, 1997;	
Velarde et al., 1997;	
Kairy et al., 1998;	
Petrosky and Farshad, 1988;	
Al-shammasi, 2001;	
Gharbi & Elsharkawy 1999	
Levitan and Murtha, 1999;	
Dindrouk and Christman, 2001	
Moradi et al., 2010	
Okoduwa and Ikiensikimama 2010	
Asoodeh and Bagheripour, 2012	
Shojaei et al., 2014	
Talebi et al., 2014	
Elmabrouk et al., 2014	
Kanu and Ikiensikimama, 2014	
Jarrahian et al 2015	

Table 2- Description of the data used to develop and test the extended model

Input	Training model (179 Data)			Testing (28 Data)			Testing (7 new experimental)		
	Min	Ave	Max	Min	Ave	Max	Min	Ave	Max
N_2	0.00	0.35	1.67	0.00	0.27	1.67	0.10	0.26	0.35
CO_2	0.00	1.27	11.37	0.00	1.17	11.37	0.14	0.41	0.83
H_2S	0.00	0.15	3.68	0.00	0.15	3.22	0.00	0.00	0.00
C_1	13.16	33.86	74.18	19.50	34.79	73.36	20.64	28.25	42.79
C_2	3.36	8.07	13.71	4.64	8.26	11.63	6.69	7.88	10.82
C_3	0.89	6.59	11.87	2.46	6.76	12.03	6.05	6.59	7.10
C_4	0.95	4.46	8.4	1.66	4.41	6.58	4.03	5.05	5.93
C_5	0.40	3.21	5.95	1.00	3.11	5.23	3.36	4.33	5.52
C_6	0.00	3.14	6.37	0.99	3.15	5.46	2.88	3.74	4.33
C_{7+}	10.72	38.89	57.73	11.18	37.88	51.22	27.61	43.49	51.31
$SG.C_{7+}$	0.74	0.87	0.959	0.70	0.86	0.93	0.85	0.89	0.92
MWC_{7+}	134	240	368.9	105	223	292	216	262	298
T, °F	58	156	319	58	165	261	121	157	230
Ps, psia	537	2198	5065	1025	2307	5000	1227	1779	3344

Table 3: Error analysis of the proposed models in comparison with all the investigated methods

Method	ARE	AARE	SD	R^2%
PR-EOS	6.24%	10.19%	1028	97.9
SRK-EOS	-3.12%	10.36%	1024	96.7
Elsharkawy (2003)	0.85%	8.57%	1015	96.4
Alquraishi (2009)	-6.06%	10.16%	1088	96.7
Bandyopadhyay (2011)	8.05%	12.28%	1054	96.8
Farasat (2013)	NA	NA	NA	NA
Gholami et al. (2014)	?	?	?	?
Ahmadi et al. (2014)	?	?	?	?
Jarrahian et al (2015)	35.52	35.56	983	95.2
New Extended Model	-0.37%	8.20%	1030	97.6
New Lumped Model	-0.50%	8.46%	1036	97.4

ARE = Average Relative Error

AARE = Average Absolute Relative Error

SD = Standard Deviation

R^2 = Correlation Coefficient

NA = Calculation was not possible, because the weight and bias were not provided.

? = The method produces unreasonable results as indicated in the appendix.

Table 4- Comparing model accuracy for calculations of saturation pressures for various crude oils

Volatile-Oil-Samples (C7+<30)																						
Fluid Source	index	N2	CO2	H2S	C1	C2	C3	C4	C5	C6	C7+	GR.C7+	MWC7+	T, °F	Ps, psi	SRK	PR	Elsh	Alq	Bandy	Jarr	Extd	Lump
Rosenegger	7	0.03	8.39	0	47.4	10.3	6.12	4.2	2.89	2.05	18.6	0.83	180	295	4000	3933	3879	4460	4420	4165	3317	4139	4178
Vogel-80	13	0.65	0.02	0	45	12.5	8.93	6.03	3.02	1.44	22.4	0.81	184	140	3002	2624	2548	2742	2895	2643	1812	3019	2996
Dansh-92	18	0	0	0	46.8	8.77	7.44	4.01	2.56	4.02	26.4	0.77	158	212	2941	2550	2594	3248	3617	3269	1944	3244	3374
	31	0.44	0.38	0	49.1	7.6	6.13	3.84	2.52	2	28	0.84	231	199	3739	3522	3507	3418	3951	3508	2309	3625	3644
Middle East	45	0.03	0.97	0	41.6	10.8	7.52	4.5	3	2.43	29.1	0.85	208	241	3069	3109	2979	3080	3513	3138	1962	3127	3140
Coats-86	51	0.55	1.03	0	36.5	9.93	8.85	6	3.78	3.56	30.4	0.84	200	234	2746	2537	2472	2714	2945	2643	1682	2777	2720

Black-Oil -Samples (35<C7+<55)																							
Fluid Source	index	N2	CO2	H2S	C1	C2	C3	C4	C5	C6	C7+	GR.C7+	MWC7+	T, °F	Ps, psi	SRK	PR	Elsh	Alq	Bandy	Jarr	Extd	Lump
Coats-86	73	1.64	0.08	0	28.4	7.16	10.5	8.4	3.82	4.05	36	0.84	252	131	1694	1674	1618	1615	1738	1644	1084	1713	1713
Middle East	88	0.08	1.37	0.69	36	8.67	5.94	2.97	1.5	3.1	39.7	0.9	274	168	2505	2710	2318	2247	2608	2321	1476	2441	2445
Middle East	114	0.29	0.48	0	28.4	8.29	7.38	5.06	3.42	4.41	42.3	0.88	252	133	1632	1672	1477	1544	1748	1517	984	1587	1598
Middle East	122	0	1.25	0	33.4	9.24	6.07	2.4	1.62	2.78	43.3	0.85	252	230	2110	2313	2278	2424	2737	2438	1410	2388	2412
Middle East	134	0	0.17	0	28.6	8.63	7.43	3.81	2.61	4.11	44.7	0.86	249	134	1548	1551	1429	1477	1764	1433	937	1503	1567
Middle East	146	0.44	0.83	0	27.8	7.54	6.9	4.61	3.1	3.11	45.7	0.9	272	133	1565	1799	1505	1558	1730	1528	1002	1602	1590
moharam-96	149	0.21	0.34	0	20	7.93	8	6.6	5.87	5.08	45.9	0.86	230	235	900	1353	1290	1531	1514	1225	773	1303	1294
Middle East	162	0.35	0.47	0	26.5	7.71	6.05	4.03	4.08	4.22	46.6	0.89	253	135	1500	1613	1380	1427	1631	1391	923	1478	1456
Middle East	176	0.12	0.51	0	28.8	7.91	6.46	3.17	2.07	3.25	47.7	0.88	250	134	1615	1665	1476	1574	1783	1502	981	1571	1614
Middle East	194	0.21	0.15	0	27.8	7.68	6.19	3.41	1.81	2.84	49.9	0.86	228	134	1590	1532	1377	1545	1680	1426	923	1466	1493
Rosenegger	207	0.37	0.02	0	17.3	6.11	8.23	5.47	3.76	3.96	54.8	0.85	242	156	537	938	888	914	1064	836	558	661	731

Newly-Measured Kuwait University Laboratory																						
Fluid Source	index	N2	CO2	H2S	C1	C2	C3	C4	C5	C6	C7+	GR.C7+	MWC7+	T, °F	Ps, psi	SRK	PR	Elsh	Alq	Bandy	Jarr	Extd	Lump
KU-1	209	0.33	0.22	0	25.6	6.87	6.39	5.61	4.68	4.33	46	0.879	222	133	1595	1360	1530	1442	1514	1394	860	1357	1337
KU-2	210	0.35	0.47	0	26.5	7.71	6.05	4.03	4.08	4.22	46.6	0.893	253	135	1500	1400	1620	1427	1631	1391	924	1478	1456
KU-3	211	0.27	0.74	0	23.1	7.2	6.45	5.93	4.66	4.16	47.5	0.906	290	138	1283	1140	1360	1284	1448	1296	800	1287	1219
KU-4	212	0.22	0.26	0	20.6	6.69	6.65	5.11	5.52	3.6	51.3	0.919	298	171	1227	1120	1300	1188	1415	1170	742	1176	1131
KU-5	213	0.19	0.14	0	27.8	8.37	6.88	5.28	4.33	3.85	43.2	0.89	278	121	1524	1300	1560	1402	1659	1409	928	1534	1489
KU-6	214	0.33	0.24	0	31.3	7.48	6.58	4.86	3.7	3.17	42.3	0.9	274	171	1978	1800	2200	1975	2242	1985	1224	2060	2026
KU-7	215	0.1	0.83	0	42.8	10.8	7.1	4.51	3.36	2.88	27.6	0.85	216	230	3344	3070	3400	3083	3568	3139	2003	3213	3200

Low-Temperature Oil Samples (59°F<T<71°F)																						
Fluid Source	index	N2	CO2	H2S	C1	C2	C3	C4	C5	C6	C7+	GR.C7+	MWC7+	T, °F	Ps, psi	SRK	PR	Elsh	Alq	Bandy	Jarr	Extd	Lump
Middle East	101	0	1.12	0.21	27	10.8	9.23	4.97	2.62	3.33	40.8	0.88	249	71	1247	1090	1033	1040	1235	564	808	1148	1205
Middle East	117	0.36	0.17	0	27.2	8.93	8.6	6.07	3.71	2.55	42.4	0.88	271	68	1130	1028	1245	1094	1248	766	820	1239	1228
Middle East	135	0	0.17	0	28.6	8.63	7.43	3.81	2.61	4.11	44.7	0.86	249	70	1270	1055	1255	1130	1311	783	826	1229	1297
Middle East	138	0.14	0.54	0	29.4	7.96	6.19	3.96	3.25	3.84	44.7	0.9	290	59	1265	1065	1350	1191	1301	880	894	1408	1390
Middle East	147	0.44	0.83	0	27.8	7.54	6.9	4.61	3.1	3.11	45.7	0.9	272	67	1320	1125	1400	1201	1268	921	879	1319	1312
Middle East	173	0.06	0.81	0	31.3	7.42	5.68	2.82	1.63	2.64	47.6	0.89	270	68	1460	1215	1535	1441	1463	1026	975	1548	1576
Middle East	193	0.16	0.3	0	24.7	7.51	6.92	4.76	2.65	3.43	49.6	0.88	239	71	1050	975	1156	1008	1107	701	714	958	986

Elsh: Elsharkawy (2003) Alq: AlQuraishi (2009) Band: Bandyopadhyay and Sharma (2011) Jarr: Jarrahian et. al (2015)
Extd: Extended model Lump: Lumped model

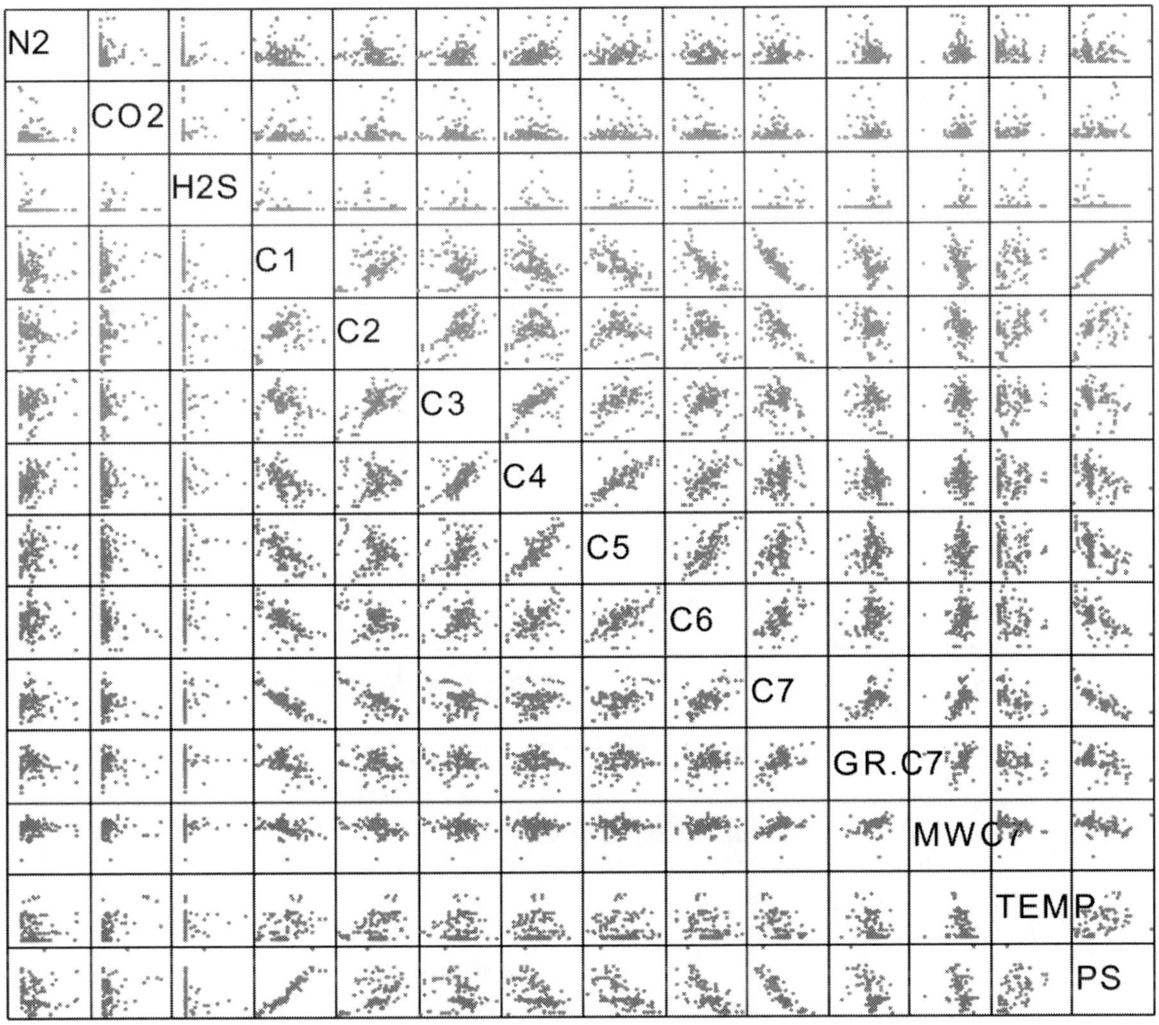

Figure 1: Correlation matrix for all the variables in the oil data bank

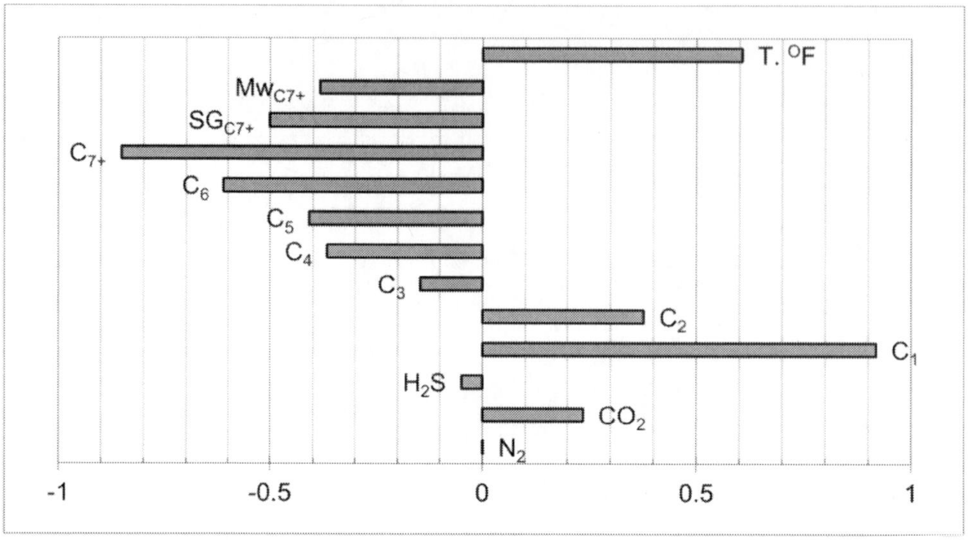

Figure 2: Correlation coefficient for all the variables in the training model (179 data)

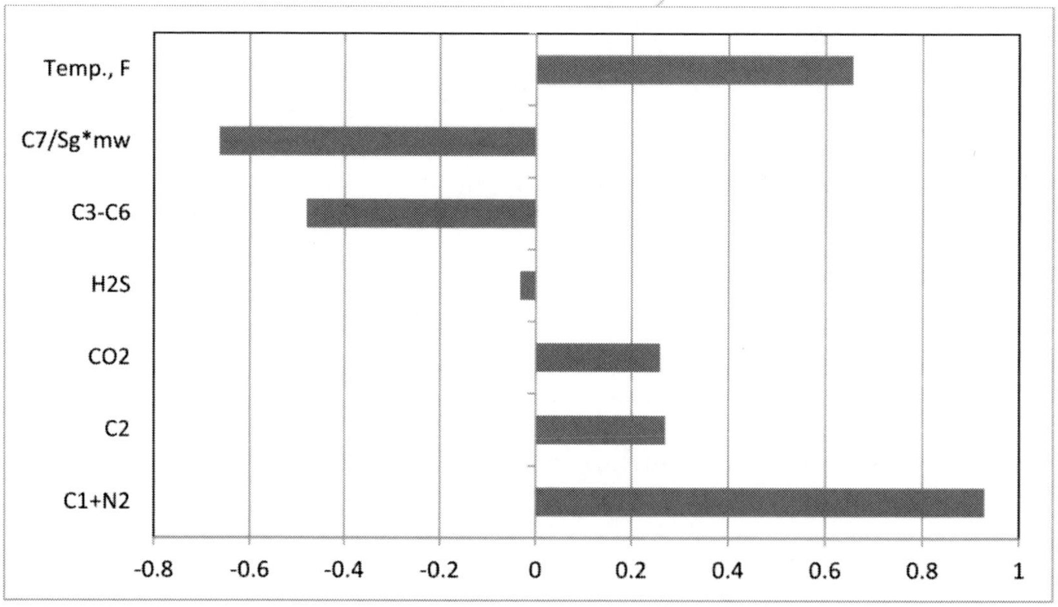

Figure 3: Correlation matrix for the lumped model (179 data)

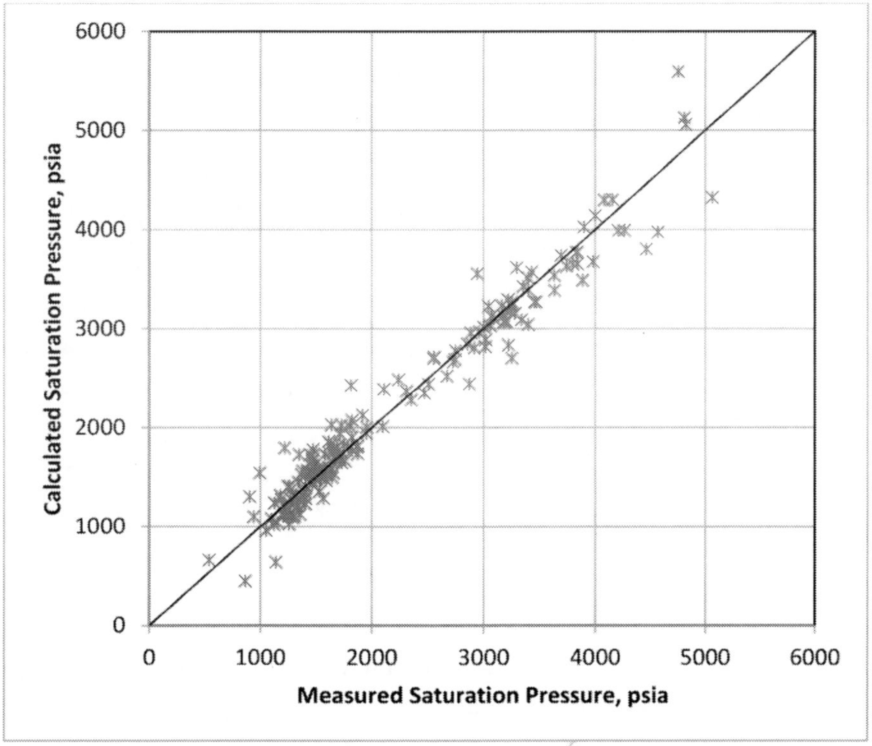

Figure 4A: Cross plot of the extended model using 179 training data

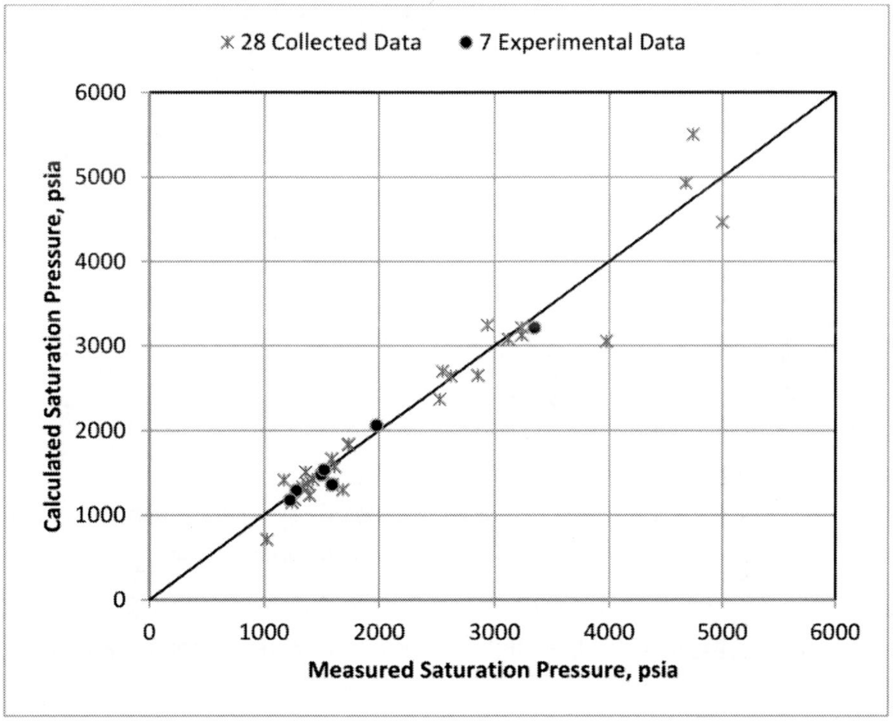

Figure 4B: Cross-plot of the extended model using 35-testing data

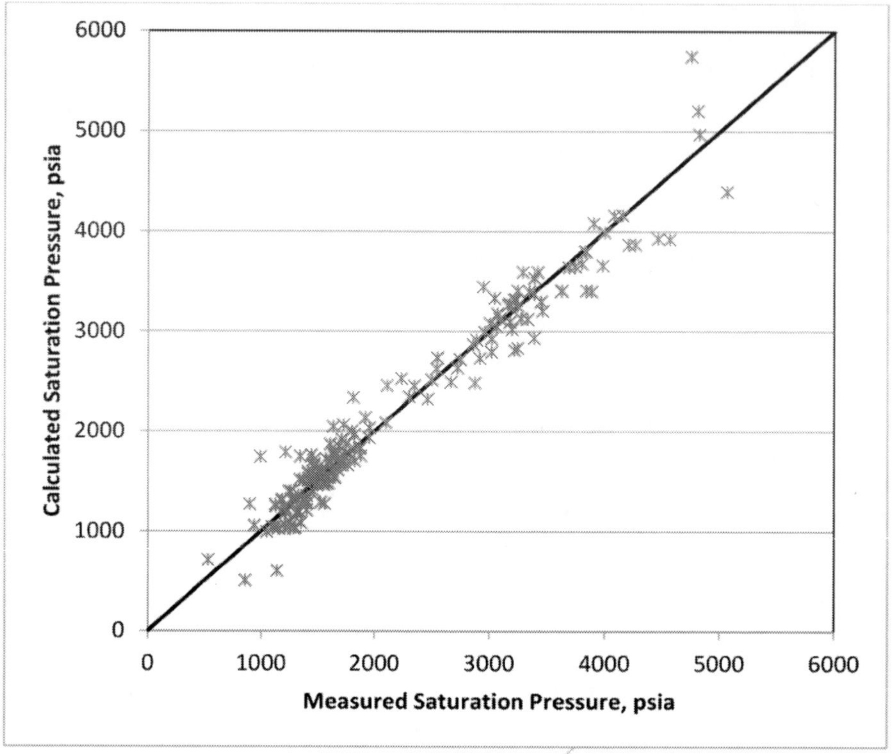

Figure 5A: Cross-plot of the lumped model using 179 training data

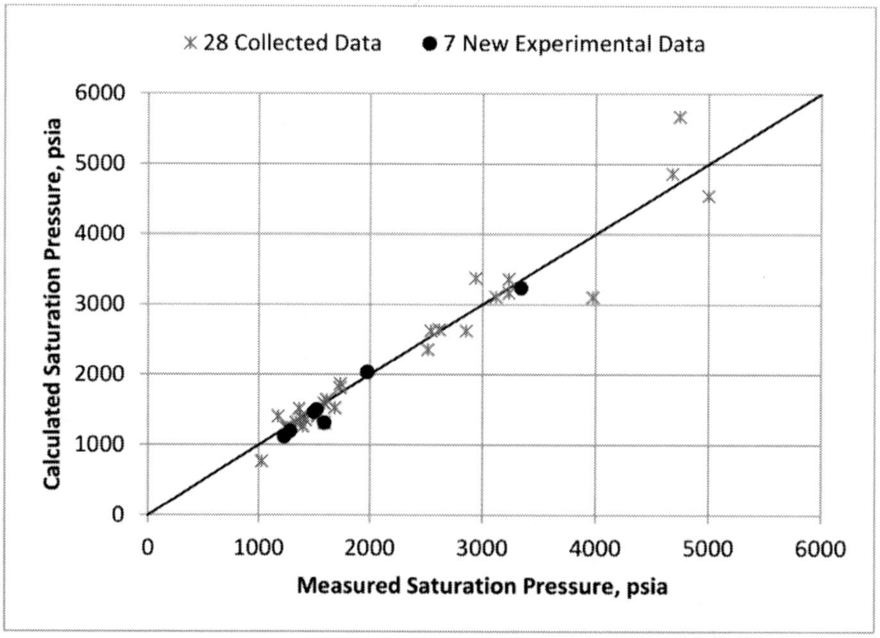

Figure 5B: Cross-plot of testing data for the lumped model

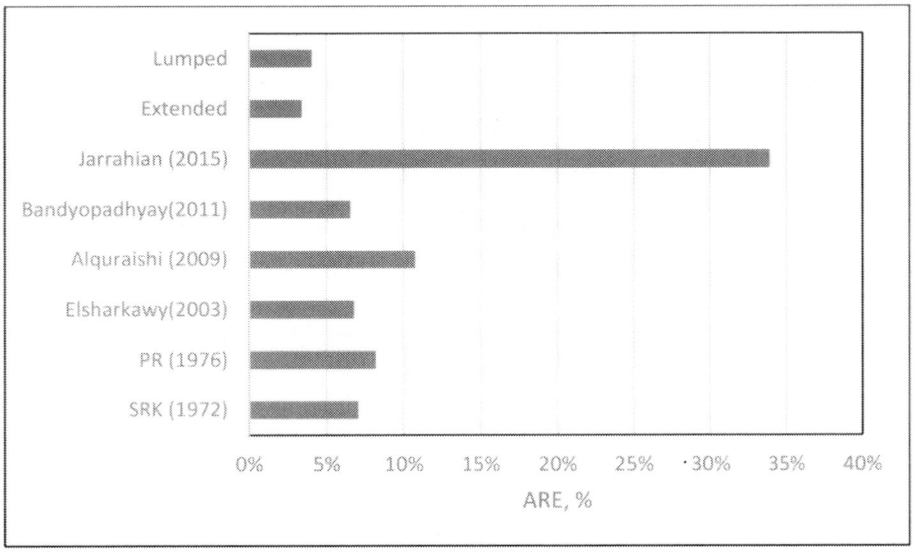

Figure 6A: Comparison of all the models' accuracy for the volatile oil samples

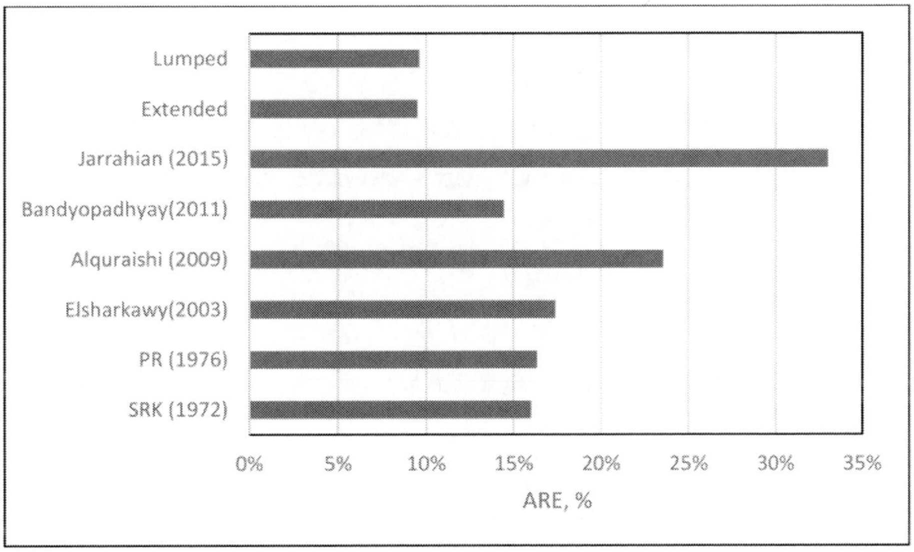

Figure 6B: Comparison of all the models' accuracy for the black oil samples

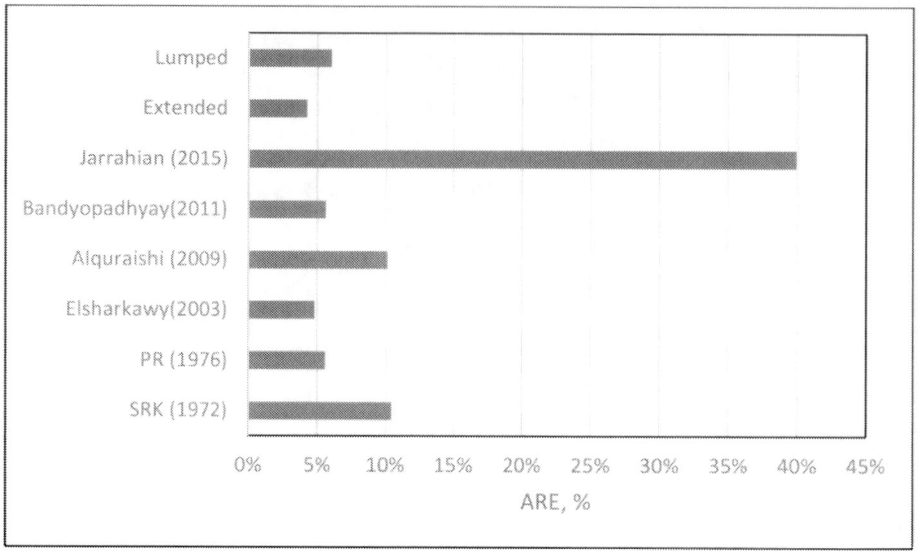

Figure 6C: Comparison of all the models' accuracy for the newly measured samples at KU

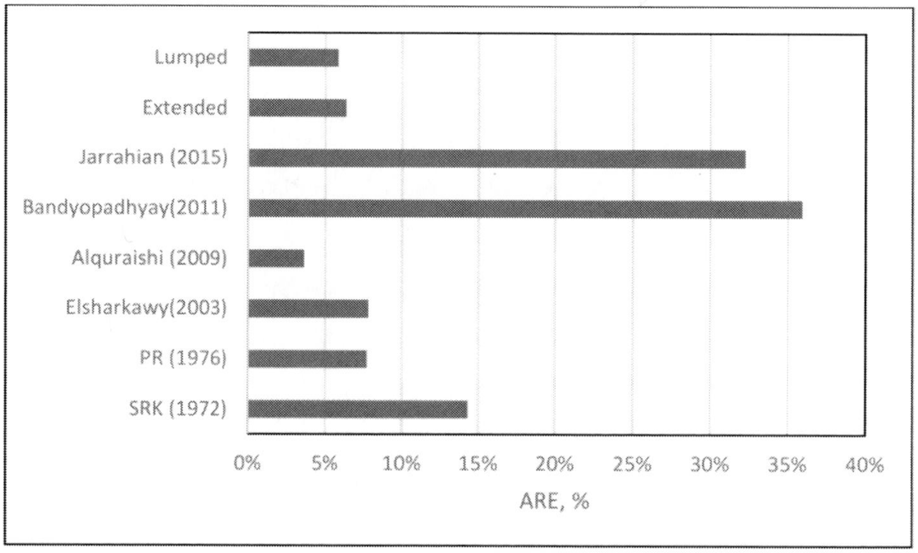

Figure 6D: Comparison of all the models' accuracy for low-temperature oil samples

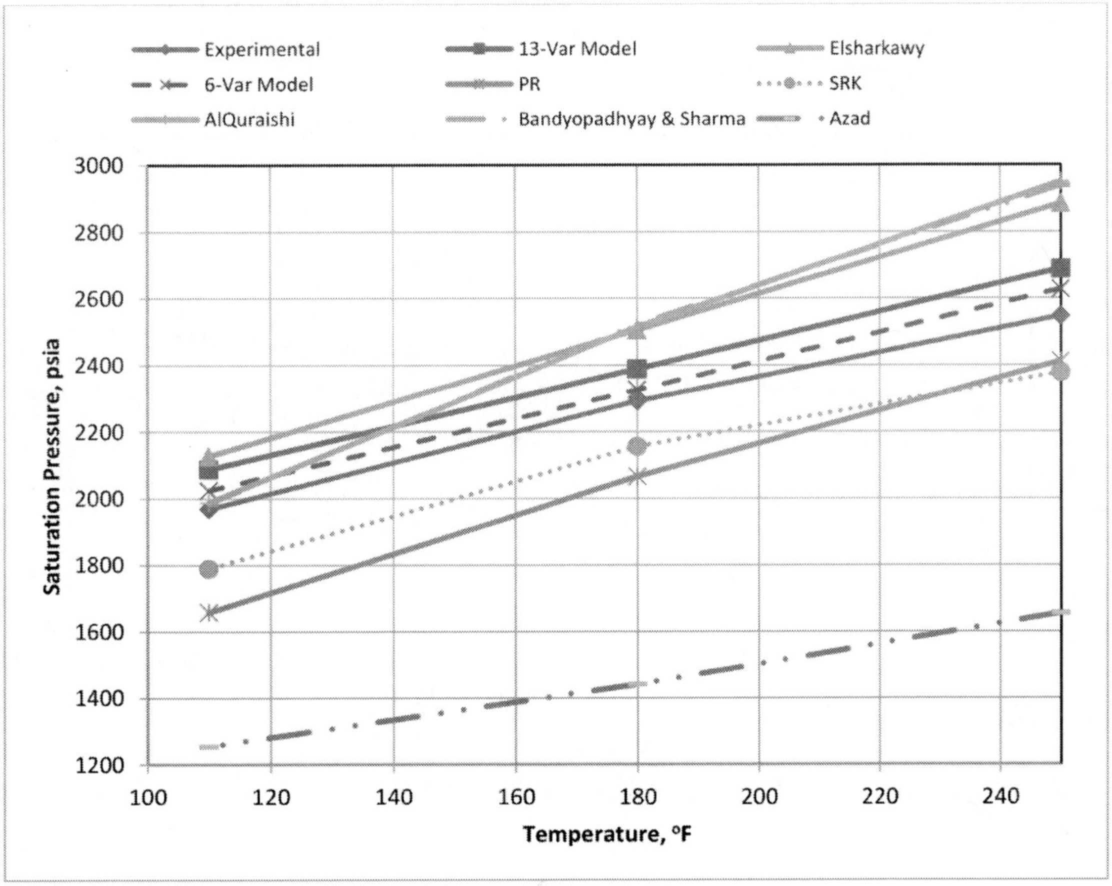

Figure 7: Changes in saturation pressure as a function of temperature for the different models (Sample #70)

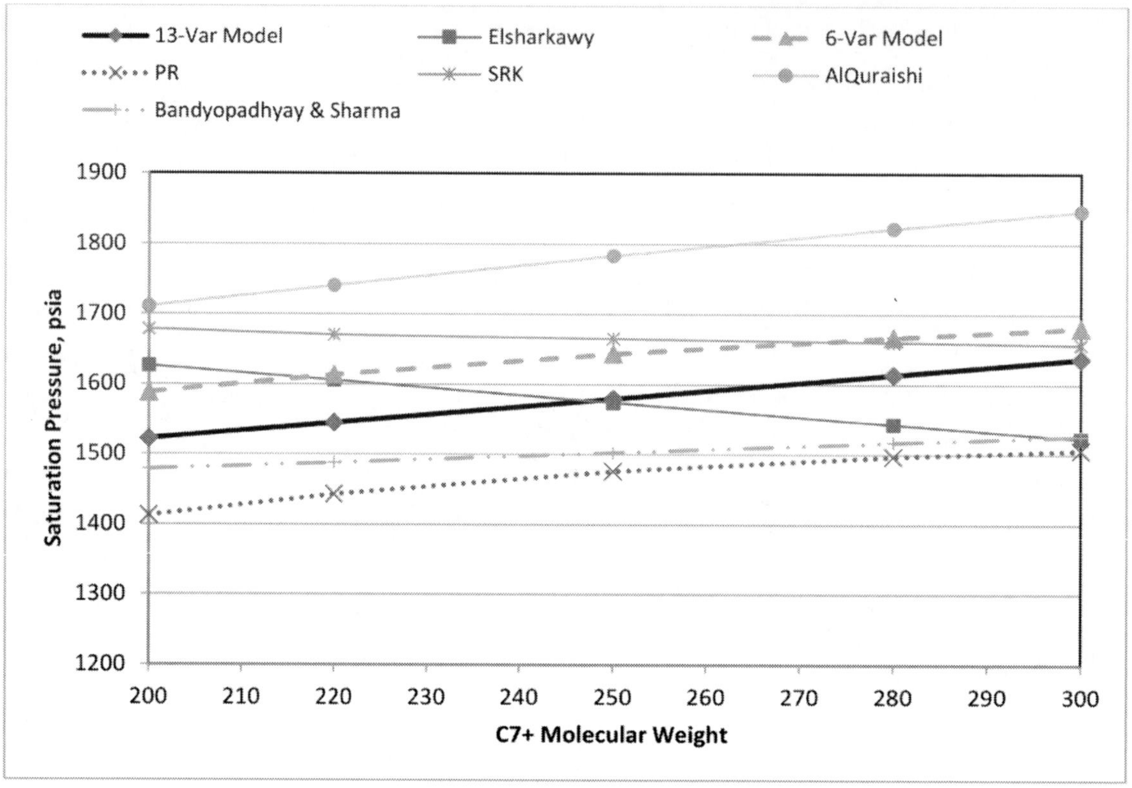

Figure 8: Comparing model performances: Changes in BPP as a function of C_{7+} molecular weight at 134°F (Sample #176)

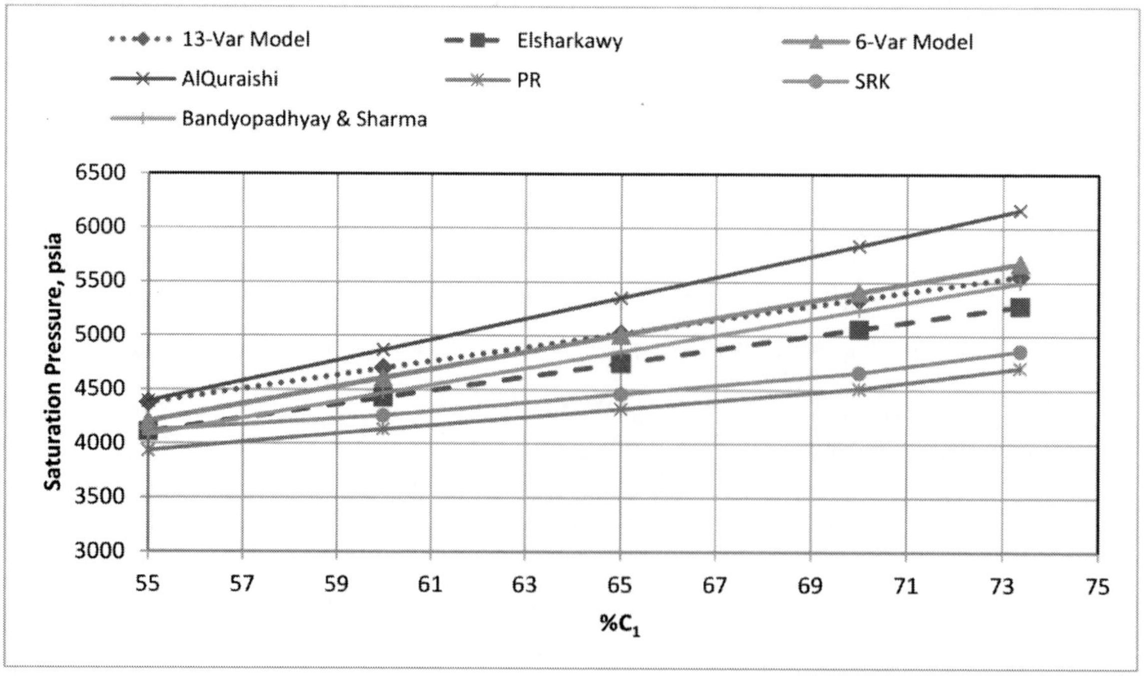

Figure 9: Sensitivity of saturation pressure to methane content at 212°F (sample #2)

439811 Purification of Ibuprofen Enantiomers By a New Technique Combining Distillation and Crystallization

Abstract:

Stripping crystallization (SC) is a new separation technique which combines distillation and crystallization. Basically, SC is operated at a triple-point condition, in which the liquid is simultaneously vaporized and crystallized due to the three-phase equilibrium. In essence, the process is continued until the liquid phase is completely eliminated and only the pure solid crystals remain in the feed. Thus, crystal washing is not required since only pure solid crystals remain in the feed and no impurities are adhered on the crystal surfaces at the end of SC. In the previous research, SC has been successfully applied to separate p-xylene from the mixed xylenes by Shiau et al. (2005, 2008, 2013).

Ibuprofen, also called 2-(isobutylphenyl)-propionic acid, is a non-steroidal anti-inflammatory drug widely used to treat headaches and minor pains. Although it is known that S-ibuprofen is responsible for the anti-inflammatory effects while R-ibuprofen might cause some long-term side effects, ibuprofen is still sold as a racemic mixture in the market due to the complexity involved in the separation of enantiomers. A thermodynamic model is developed in this work to simulate the three-phase equilibrium conditions during the SC operation. The experiments show that seeding with ultrasound mixing or magnetic stirring can be efficiently employed to purify S-ibuprofen from ibuprofen enantiomers. The experimental results, including the final enantiomeric purity and recovery ratio of S-ibuprofen, are consistent with the simulation results predicted by the model.

References:

1. Shiau, L. D.*, Wen, C. C. and Lin, B. S. "Separation and purification of p-xylene from the mixture of m-xylene and p-xylene by distillative freezing", *Ind. Eng. Chem. Res*, **44(7)**, p2258-2265, 2005.

2. Shiau, L. D.*, Wen, C. C. and Lin, B. S. "Separation of p-xylene from the multicomponent xylene system by stripping crystallization", *AIChE Journal*, **54(1)**, p337-342, 2008.

3. Shiau, L. D.*; Liu, K. F. "Investigations into the effects of the cooling rate on stripping crystallization", *Ind. Eng. Chem. Res*, **52**, p1716-1722, 2013.

441535 Hydrogen Disposal Flare System Hazard Identification and Mitigation Analysis

Abstract:

A hazard assessment on a hydrogen disposal flare system was conducted in preparation for a hydrogen turbopump test at the Air Force Research Laboratory (AFRL), Aerospace Systems Directorate at Edwards AFB, CA. The test operations required the disposal of large quantities of liquid and gaseous hydrogen through a commercial flare system to prevent the formation of an explosive environment. As part of the assessment, the primary steps in the test operation process were evaluated for the formation of a potential hazard by looking at each associated failure mode and the resulting affect on personnel and facilities. Evaluation of each operational step was conducted in a Process Failure Modes and Effects Analysis (PFMEA) that identified the probability and severity of the identified effects and associated hazards. If the probability and severity were unacceptable, mitigating steps were applied until the overall hazard risks were reduced to acceptable levels. The recommendations of this assessment were incorporated into both the flare system design and the operational procedures. This paper outlines the results of the PFMEA which include identification of the effects/hazards, available risk-mitigating techniques, and implementation of these techniques used for the mitigation of the identified hazards. Based upon the findings significant design modification to the vender's commercial equipment was needed to comply with recent changes in the code. Most of the identified hydrogen flare system hazards/effects were categorized into four groups: internal explosions, obstructions, external explosions, and maintenance degradation. Many corrective actions were implemented to mitigate both the probability and severity of the identified hazards. In addition, this paper discusses the installation of oxygen monitoring systems on the flare system, determination of appropriate system purge requirements, installation of purge system status alarms, incorporation of a triple redundant flare ignition system with status indicators, and a number of other hazard risk mitigation techniques. This comprehensive description of the hazards and mitigation of a hydrogen disposal system is a plant is vital part to safe operation of both hydrogen commercial and aerospace facilities.
<align="center">

442068 Enhanced Furfural Production By an Innovative Bottom Dividing Wall Column with a Decanter Configuration

Abstract:

Enhanced Furfural Production by an Innovative Bottom Dividing Wall Column with a Decanter Configuration

Nguyen Van Duc Long, Le Cao Nhien, Le Quang Minh and Moonyong Lee[†]

School of Chemical Engineering, Yeungnam University, Gyeongsan 712-749, South Korea

Furfural, which are used to derive other industrial chemicals, has been identified as one of the major bio-based platform chemicals that can compete with petroleum-based chemicals. However, the current commercial furfural process has low yield and requires intensive energy. This paper reports the simulation, detailed design and optimization of an industrial process recovering furfural from lignocellulosic biomass, a renewable and inexpensive feedstock. The aqueous feed mixture from the acidic hydrolysis of biomass was fed to the separation process to recover furfural as the main product, along with methanol as valuable byproducts. The purification process was intensified by applying an innovative bottom dividing wall column with a decanter configuration (BDWC-D). The response surface methodology (RSM), which allows the interactions between variables to be identified and quantified, was used to optimize the BDWC-D structure. The predictions by the RSM showed satisfactory agreement with the rigorous simulation results. The heat pump technology and heat integration were then employed to improve the purification process. The results showed that the proposed sequence could save the overall energy requirements and total annual cost significantly as compared to the conventional sequence. This sequence can be employed for both grass-root and retrofit designs.

Keywords: Lignocellulosic Biomass; Furfural; Energy Efficiency Improvement; Dividing Wall Column; Response Surface Methodology

Acknowledgments

The authors wish to acknowledge the financial support from the R&D Convergence Program of NST (National Research Council of Science & Technology) of Republic of Korea and KITECH (Korea Institute of Industrial Technology) (ES150001).

442557 Flow Field Simulation and Distribution Performance Analysis of a Gas Pipe Distributor in Fccu

Abstract:
A fluidized catalytic cracking (FCC) is an important petroleum refinery process, which, aided by highly selective zeolite catalysts, cracks large-molecule low-value heavy oil feedstock into small-molecule highvalue
products. When spend catalyst is fed into the regenerator for reactivating, it is fluidized by the gas distributor at the bottom of regenerator. The gas distributor is designed to uniformly distribute gas to maintain the fluidized state of the fluidized bed and to provide the reaction gas to guarantee the burning coke deposited on the spend catalyst. Therefore, the gas distribution performance of the gas distributor influences the quality of initial gas distribution directly. So the pipe distributor used in a industrial FCC unit, which is composed by main pipes and a number of branch pipes with many nozzle tubes, was investigated by means of flow field calculation. The gas distribution performance of the pipe distributor was analyzed based on the simulated flow field. The simulation result shows that there are a obvious nununiform
gas distribution from the different nozzle gas flow rate. The static pressure increases gradually along the flow direction in the nozzle tube while the axial flow rate decreases, which leads to the nozzle gas flow rate increasing gradually. Furthermore, an obvious bias flow occurs at the inlet of the nozzle tube, which influences the gas flowing into the nozzle. The pressure drops performance of nozzle and pressure varying in the branch pipe are analyzed comprehensively. Due to the nonuniform
gas distribution by the gas pipe distributor in the practical usage, the problems such as overlarge temperature difference among the fluidized bed layers in the regenerator have often happened, which will not only reduce burning coke efficiency and but also decrease activity of the regenerated catalyst. Meanwhile, the erosion of the distributor also occurs, and in consequence the function of the distributor fails, which even causes the nonscheduled shutdown of the FCC unit and leads to tremendous economic loss. The support is provided for optimizations of both structural parameters and operation parameters of the gas pipe distributor and improvement of the gas distribution performance. Furthermore, improvement suggestions were provided based on the above reasoning. The results are expected to be useful for the distributor design.

442925 Identifying Gas Liquid Contactor Performances for Optimized Natural Gas Desacidification Amines Units - from Process Calculations to Experimental Strategy

Abstract:

Amine units are widely used for gas sweetening. The absorber performances of these reactive absorption processes are typically mass transfer limited. The proper selection of gas liquid contactor is thus critical to optimize the size of towers, the solvent flowrate and the duty of the reboiler. This work proposes an approach to identify the most suitable packings for a given application.

The case-study selected here to illustrate the approach deals specifically with a CO_2removal application where the sour gas is at 7 % vol, and the sweet gas specification is 2% vol. The gas flowrate is fixed at 0,18 MMSCFD with a solvent flowrate fixed at the minimum of the thermodynamic constraints (via equilibrium curves). An in-house rate-based simulator with accessible forms of mass transfer correlations, based on a double film model, is used to calculate the absorber height and diameter which are needed to reach the sweet gas specifications. A reference structured packing is compared with standard valve trays.

The calculations show that the bed height required with the first contactor is considerably higher than with the second one while the column diameter is reduced. In this case the use of standard valve trays is more attractive for cost optimization. Interestingly, because mass transfer is accelerated, the absorption rate with the structured packing is found to be thermodynamically limited due to a more severe temperature increase which reduces the driving force in the bottom section of the absorber.

The study determined that the optimum design for the investigated case is a compromise between a full trayed tower, that generally imposes a larger diameter, and a packed tower where the higher mass transfer forces result in a temperature rise in the bottom of the tower which in turn imposes a height increase and a much higher solvent flowrate. . To determine the mass transfer characteristics of a gas-liquid contactor providing this optimum, a sensitivity analysis has been conducted. Mass transfer parameters, i.e., liquid and gas side mass transfer coefficients and interfacial area, were varied by a factor 2 relative to the ones obtained with the reference structured packing. Calculations show that one should use a contactor that reduces liquid phase resistance, specifically, the mass transfer conductance $k_L a_e$ should be increased by at least 50%. This target could be reduced to +30% if one could simultaneously reduce the gas phase conductance $k_G a_e$ by 20%. This result can be explained by the link between mass and heat transfer (Chilton-Colburn analogy) whereby reducing $k_G a_e$decreases the temperature level which in turn thermodynamically increases the mass transfer driving force.

The next step will be to review the literature more extensively to make a short list of pre-selected random and structured packings that best meet the targets. Experimental tests will then be run on IFPEN facilities to measure flooding limits and mass transfer parameters in a small 150 mm diameter column. This step is required to confirm packing's behavior, considering the disparity of the values reported by different authors in the literature. The last step is to consolidate the existing correlations for the most promising packing, and mitigate differences between various authors by performing characterization tests on larger diameter test columns. The final correlations will be implemented in our in-house rate based simulator for industrial use.

The present approach can be applied to determine the required mass transfer characteristics for optimizing other gas treatment applications. However, the conclusions reached here as regards to mass transfer targets and the problematic temperature levels, are valid only to the present case –study, and are not directly applicable to other gas sweetening applications (selectivity, LNG, ...). For example, in a case where H_2S and CO_2 are treated simultaneously the mass transfer limitations could be very different.

*** Formatting Issue - Best Available Paper/Graphic ***

443365 Technical Reconstruction of Double Effect Distillation for Methanol Process

Abstract:

Abstract: Methanol is one of important basis raw materials of chemical industry and also an excellent environment-friendly fuel, which make it as a significant commodity in the global market. Distillation is an essential unit operation for purified production of methanol. Product quality, production cost, environmental impact and energy consumption are largely related to distillation process.

Our research effort, under this circumstance, has been devoted since 2002 to energetically optimize 4-column methanol distillation system that is widely adopted in methanol units from coal, with reduced cycling water requirement. This patented process technology saw its industrial application in Chinese largest single unit of 1800Kt/a of Baofeng Group located at North China, with the lowest steam consumption of 0.9/1 t(steam)/t(methanol) compared with normally 25% higher for both simulative case study and on-site operation. Methanol product is in accordance with O-M-232G and the recycling water is reduced significantly from 67 t(CW)/t(methanol) to 52 t(CW)/t(methanol).

The flow diagram of this advanced double effect distillation process is showed in Fig.1. Its advantages arise for limitations found in current units: (1) heat exchange network failed to fully recover energy causing large CW flow to remove extra heat; (2) for pressured column, bubble point reflux need larger reflux ratio than the sub-cooled; (3) the heat of distillate from pressured column was normally ignored; (4) the low concentration of methanol of feed inlet to recovery column leads to higher reflux radio, responsible for unsatisfied unit energy efficiency.

Fig.1 The double effect distillation process.

(Column1.Light ends column; Column 2.Pressured column;

Column 3.Atmospheric column; Column4.Recovery column)

As a solution, a new flow diagram is built up and showed in Fig.2. (1) HEN totally reorganized (not completely show); (2) sub-cooled reflux to pressured column caused 10% reduction to the reflux ratio, economizing steam consumption; (3) top product of pressured column give heat to crude methanol before feeding light ends column; (4) change product and by-product position of recovery column to give efficient unit energy consumption.

Fig.2 The reconstruction of distillation process.

(Column1.Light ends column; Column 2.Pressured column;

Column 3.Atmospheric column; Column4.Recovery column)

It is estimated that the production of methanol will be over 60 million tons in 2014. If this reconstruction could be applied in the whole methanol industry, 1.3 million tons of standard coal would have been saved. This is equivalent to reduce 5 million tons emission of carbon dioxide, so the environmental impact can be largely decreased. Field operation shows reconstruction perfect substitution worth to spread to the whole methanol industry, energetically efficient and low environmentally impacting.

443095 Experimental and Numerical Study of Transport Phenomenon of Falling Film Flow with Different Heating Methods

Abstract:

Falling film flow is a common phenomenon in chemical industry. It is widely used in chemical instruments, such as structured packed columns and falling film evaporators. An experiment was implemented to study the mass transfer phenomenon of binary-mixture (ethanol-water) falling film flows with different heating methods (uniform or concentrated heating). In order to study the detailed flow and mass transfer behaviors, a 2-D multiphase computational fluid dynamics (CFD) model based on volume of fluid (VOF) was developed. This model divides mass transfer into two parts: one is caused by the vapor-liquid equilibrium (VLE) and the other is caused by the heat flux. Results from simulation are inconsistent with the experimental data. Both experiments and simulations show that the ethanol concentration of the vapor with concentrated heating is 1.5% higher than the one with uniform heating.

443101 Thermodynamic Modeling of HNO_3-H_2so_4-H_2o Ternary System with Symmetric Electrolyte NRTL Model

Abstract:

Nitric acid concentration/sulfuric acid concentration (NAC/SAC) process has been widely used for concentrating dilute aqueous nitric acid and recovering spent sulfuric acid. Nitric acid is concentrated by employing sulfuric acid to bind water and break nitric acid-water azeotrope at approximately 68 wt. % nitric acid. To support heat and mass balance calculations and process simulation for NAC/SAC processes, we develop a comprehensive thermodynamic model for nitric acid-sulfuric acid-water ternary system based on previously published thermodynamic models of nitric acid-water [Wang et al., Fluid Phase Equilibria, (2015) (DOI: 10.1016/j.fluid.2015.04.015)] binary and sulfuric acid-water [Que et al., Journal of Chemical & Engineering Data, 56 (2011) 963-9772] binary systems developed with symmetric electrolyte NRTL (eNRTL) activity coefficient model. The present work takes into account one-step hydration of nitric acid molecules, two-step hydration of sulfuric acid molecules, and further hydration of hydronium ions. With up to three temperature coefficients for each eNRTL binary interaction parameter and up to two temperature coefficients for each chemical equilibrium constant, the ternary model is validated with boiling point data from 383.15 to 471.45 K and activities at 273 K. Furthermore, contour plots of boiling points, vapor phase composition, and specific heat capacity of the ternary system are generated for engineering use.

Keywords:

Nitric acid, Sulfuric acid, Electrolyte NRTL model, Vapor-liquid equilibrium, Boiling point, Heat capacity, Nitric acid concentration/sulfuric acid concentration (NAC/SAC)

443503 A Computational Analysis of Fire Scene for Safety in a Compartment

Abstract:

This article details numerical prediction of fire spread carried out to study the distribution of carbon monoxide concentration, smoke dispersion, soot concentration and temperature in laboratory compartment, using a commercial fire modeling software tool. The heat release rate curve predicted from the simulation was compared with literature values carried out in a road tunnel. The time variations of carbon monoxide transportation, smoke concentration and temperature spread in compartment room are reported to evaluate the effect on humans. The soot concentration was described and used throughout this study as an indicator of safety in the compartment room. The effects of vent dimension and ventilation capacity were analyzed by smoke detector response times. This work provides improvements derived from reinterpretation of actual fire scene considering the ability of mechanical exhaust system and ventilation in compartment room to mitigate the hazard of fire.

443640 Preparation and Characterization of (*FURCRAEA spp*) as Green Support of Iron Nanostructured Catalyst

Abstract:

PREPARATION AND CHARACTERIZATION OF (*FURCRAEA spp*) AS "GREEN" SUPPORT OF IRON NANOSTRUCTURED CATALYST

Karen G. Bastidas G[1], Anamaria Barrera B[1], Cesar A. Sierra A[2], Hugo R. Zea R[3]

[1] Department of Chemical and Environmental Engineering. National University of Colombia (Bogota).

[2] Associate professor. Director of the Research Group of Macromolecules. National University of Colombia (Bogota).

[3] Associate professor. Director of the Research Group of Materials, Catalysis and Environment. National University of Colombia (Bogotá).

Abstract.

The availability of natural water resources has declined over time, becoming one of the biggest problems today. Environmental risk in Colombia's rivers has increase due to the widespread use of harmful chemicals in a variety of processes, ranging from illegal mineral extraction to industry wastewater, e.g. Medellin's river, due to its strategic location along a textile industrial corridor might be subject to dye contamination; the gold and platinum mining zones in Choco and Antioquia regions are being heavily affected by the use of mercury in the extraction processes. In 2009, 520 tons of mercury were imported by miners working in this area, amount that exceeds several times the mercury used by regulated industry in previous years (Ministry of Mines and Energy of Colombia, 2007). Not only the environmental risk is a concern but also the economic conditions of many low-income families whose livelihood in many cases depends exclusively of the of the mining.

Due to the issues raised above, the main goal of this study is to research, develop and implement innovative technologies in the synthesis of highly efficient, low cost and environmentally friendly iron based catalysts capable of degradate pollutants, ensuring operational stability. A native fiber (Fique, FURCRAEA spp) has been selected as a solid support for these catalysts due to its abundance, low cost and special physical and chemical characteristics.

The combination of cellulosic fibers with iron oxide nanoparticles could provide exceptional biodegradable composite materials for the treatment of organic dyes and heavy metals present in wastewater. Likewise, the use of iron particles immobilized on fique fibers allows an easy separation catalyst-treated water, turning the biocatalyst in highly recyclable and reusable material.

Fique fiber was initially characterized by determining the percentage of ashes, total lignin, soluble and insoluble lignin, cellulose, hemicellulose, total solids and humidity, obtaining 0.94, 22.58, 0.5, 22.8, 36.3, 27.5 0.701 and 2.1%, respectively. Additionally, the isoelectric point (pzc), acidic and basic sites were measured.

Then, the fiber was subjected to a pretreatment in order to expose more active sites which on time serve as nucleation points for the deposition of nanoparticles. This was done by means of an ultrasound-assisted procedure, varying the sonication time (30, 60 and 90 minutes). The effect of the sonication was followed by measuring the contents of lignin, cellulose and hemicellulose within the fiber after each time. 60 minutes of sonication was found to be most beneficial time since the amount of lignin and hemicellulose present in the fiber was reduced, while conversely the amount of crystalline cellulose exposed increases.

In order to improve the chemical ability of the nucleation sites to interact with iron and then to form stable and size controllable nanoparticles, a cationization procedure was performed on the fiber; this was achieved by the exposure of the fiber to acidic and basic conditions promoting cationization. Three cationization times (1, 2 and 3 hours) were studied and their effect on the fiber chemical surface was evaluated using FTIR-ATR and XRD analysis. The results showed 3 hours as the optimal cationizing time.

Subsequent to the cationization of the fiber surface, a procedure of ionic/cationic exchange was implemented by exposing the fiber to a solution of iron salts in order to replace the cations in the fiber surface and anchoring the iron nanoparticles. Different conditions were studied in order to synthetize zero valent iron nanoparticles (nZVI) and others with different oxidation states. Finally, to track the impregnation process on the fiber, samples from the precursor solution were taken at different times and its iron concentration was determined by atomic absorption spectrophotometry (AAS). Complementary characterization performed by X-ray diffraction, BET surface area and electronic microscopy, among others.

Key words: Natural support (fique), iron nanoparticles, bionanocatalyst.

443764 Optimization of Heat Exchanger Network in Steam Assisted Gravity Drainage Oil Sands Surface Facilities

Abstract:

Steam assisted gravity drainage (SAGD) is the most viable and environmentally safe recovery technology for extracting heavy oil and bitumen. It is a thermal in-situ production process where parallel wells are drilled horizontally into an underground bitumen reservoir. Steam is produced at the central processing facility (CPF) and injected into the reservoir through the wells. The steam heats the bitumen to a point where gravity allows it to flow down to the lower well where the mixture of bitumen and water is then pumped to the surface. The water and bitumen are separated at the CPF. Typically, more than 90% of the energy consumed in producing a barrel of bitumen using SAGD is used to generate the steam that is injected into the bitumen reservoir. Fossil fuels is employed in the process, which causes GHG (greenhouse gas) emissions. The usage of hot and cold utility is reduced by increased heat exchange and optimization of the heat exchanger network, which results in reduction of GHG emissions. In this work, pinch analysis is performed to minimize energy consumption.

443774 Thermodynamic Anaysis of Dehydration of n-Propanol Using Heterogeneous Azeotropic Distillation

Abstract:

Heterogeneous azeotropic distillation is a widely popular process in chemical process industries for the separation of binary azeotropic mixtures into their constituent pure components. It involves the addition of an entrainer which is partially miscible and forms a heterogeneous azeotrope preferably with one of the original components. This is a powerful technique which combines several physical phenomena such as presence of azeotropes and liquid-liquid immiscibilities, and is used widely in industries to enhance difficult or intractable distillation separations. For the design of heterogeneous azeotropic distillation columns, a detailed study of the heterogeneous region is crucial to plan the distillation sequence, which requires experimental vapor-liquid-liquid equilibrium (VLLE) data.

The present study focuses on the investigation of the heterogeneous region of the well known industrially available azeotropic mixture n-Propanol-Water system using Propyl acetate as entrainer. The VLE data of the system was generated by means of modified Othmer still and the still is modified using a probe type ultrasonic homogenizer to account for the heterogeneity. Isobaric vapor- liquid- equilibria and vapor - liquid - liquid - equilibria data at atmospheric pressure have been obtained for the ternary system n-Propanol-water-propyl acetate and the constituent binary systems. The equilibrium data has been correlated using NRTL, UNIQUAC and UNIFAC models and compared with the experimental data.

443776 Study of Ionic Liquids As Entrainers for the Extractive Distillation of *N*-Propanol and *N*-Propyl Acetate Binary Azeotropic Mixture and the Isobaric VLE Studies of the System

Abstract:

Distillation is a widely accepted technique for the separation of binary liquid mixtures. Separation by simple distillation is not possible for binary mixtures that form azeotropes. One of the commonly used methods to separate azeotropes is extractive distillation. It is the process of adding an entrainer that can alter the relative volatility of the mixture and thereby breaking the azeotropic point. Ionic liquids on account of their physiochemical properties as well as the recyclability have been used as a preferred replacement for the conventional entrainers. Extractive distillation with ionic liquids has the advantages of both liquid solvent (easy operation) and solid salt(high separation ability). The corrosion and dissolution problems by using inorganic salts can be rectified by these non-volatile, chemically stable, high electrical conductive molten salts.

The azeotropic system under consideration in the present study is *n*-propanol - *n*-propyl acetate mixture. These compounds which constitute the products/reactants of esterification reaction of n-propanol form a minimum boiling azeotrope at 94.7°C. *n*-propyl acetate has been included in a list of volatile acetates as an ecologically benign CFC-free solvent for aerosol sprays.

Vapour liquid equilibrium(VLE) measurements of *n*-propanol and *n*-propyl acetate are performed at atmospheric pressure in a modified Othmer still which is a continuous distillation still in which vapour phase is recirculated. The most suitable ionic liquids for shifting of azeotropic composition of the system are selected using COSMO-RS model with suitable combinations of anions and cations. The isobaric VLE of the system with [EMIM][HSO_4] is obtained and compared with the existing data. The influence of alkyl chain length on the performance of ionic liquids is also analysed with regard to the previous studies. The effect of polarity is the most important factor on the behaviour of ionic liquid solvation in alcohols. Experimental VLE data were correlated using modified UNIFAC model.

443784 Measurement and Correlation of the Mutual Solubility of Diisopropylamine + Water and Triethylamine + Water Systems at High Pressure

Abstract:

The mutual solubility data of diisopropylamine + water and triethylamine + water systems were measured at the pressure up to 100 MPa and the temperature range was 280.2~316.2 K. Experiments were performed in a high-pressure autoclave with a moveable piston. The experimental data were correlated satisfactorily with NRTL activity coefficient model and the binary parameters were obtained.

443819 Improving High Shear Rate Friction Reduction Behavior of Polyacrylamide-Based Fracturing Fluids Using Xanthan Gum

Abstract:

The main component of slickwater in fracturing low permeability formations is a high molar mass polymer such as polyacrylamide (PAM), which acts as a drag reduction agent in turbulent flow. Polyacrylamide does not degrade in the environment and also suffers from mechanical degradation at high turbulent flow rates. Five different concentrations of PAM as well as mixtures of polyacrylamide/xanthan were prepared to investigate the possibility of improving shear stability of PAM. Drag reduction measurements were performed in a flow loop and friction factors at different flow rates were calculated using rheological parameters. Based on the observations, Xanthan (XG) is not as good a friction reducer as PAM, even at three-fold higher concentration. XG slightly improved the shear stability in mixed polymer solutions at low concentrations. But at high concentrations, XG significantly improves the friction reduction behavior of PAM at nonionic and ionic environments.

443836 Oxidative Dehydrogenation of n-Butane on Carbon Catalysts

Abstract:

Butenes and butadiene, which are one of the important industrial chemical products to produce synthetic rubber , are sythesized traditionally by oxidative dehydrogenation (ODH) of n-butane over various mixed metal oxides. Metal-free carbon nanomaterials has recently received much attention by researchers as a catalyst for the ODH reaction due to its environmental acceptability, corrosion resistance, and unique surface properties. For the reason, the reaction system was modelled in this study for the sythesis of butene and butadiene. For the modelling of the reaction system, the catalytic performance of several carbon catalysts, such as graphite and activated carbon, were initially tested. Then, the reaction mechanisms and kinetics were developed based on experimental results. Lastly, the mathematical modeling of the reaction system was optimized to maximize an overall yield of the main products. This approach allowed us to develop the optimized reaction system more efficiently.

This research was respectfully supported by Engineering Development Research Center (EDRC) funded by the Ministry of Trade, Industry & Energy (MOTIE). (No.N0000990)

443855 Modeling the Thermal Behaviors of an Ultracapacitor Module

Abstract:

Ultracapacitors, also known as supercapacitors, have the potential to meet the increasing power requirements of the sustainable energy systems for the grid-scale energy storage systems and electrified vehicles. As compared to batteries, ultracapacitors offer a higher power density, higher efficiency, and longer shelf and cycle life. Because the performance of an ultracapacitor depends on temperature, it is important to calculate accurately the thermal behavior of a single ultracapacitor cell and the module composed of multiple ultracapacitor cells for the efficient and reliable systems integration of an ultracapacitor in grid and transport applications.

In this work, modeling is performed to study the effect of the layout configugation of the ultracapacitor cells on the thermal behaviors of an ultracapacitor module. The ultracapacitor module is subject to the charge/discharge cycling with constant-current between the nominal voltage of the ultracapacitor module and the half of this voltage. The charge/discharge current values examined are 50, 100, 150, and 200 A. The validation of the modeling approach is provided through the comparison of the modeling results with the experimental measurements.

443856 Modeling the Electrolyte Flow Behaviors of a Zinc/Bromine Flow Battery Module

Abstract:

The zinc/bromine (Zn/Br_2) flow battery is a promising rechargeable power source for energy storage systems because of its inherent chemical simplicity, high degree of reversibility at the electrodes, good energy density, and abundant low-cost materials. The Zn/Br_2 flow battery system, intended for the grid-scale energy storage system, is to be made in modular form. A Zn/Br_2 flow battery module is composed of the multiple stacks consisting of bipolar carbon electrodes. Since the stacks require electrolyte circulation on both the negative and positive sides, two plumbing loops are provided. The flow loops contain an electrolyte storage tank and a circulation pump. A heat exchanger may be included for temperature control in the flow loop. The design of the electrolyte circulation plumbing determines the distributions of electrolyte flow and pressure within the stacks and the efficiency of the stack is dependent on the flow distribution throughout the stack. The modeling of the distributions of electrolyte flow and pressure within the module is needed to optimize the plumbing for electrolyte circulation and thereby to improve the performance of the module. In this work, modeling is performed to study the effect of the layout configuration of the plumbing for electrolyte circulation on the distributions of electrolyte flow and pressure within a 50 kWh Zn/Br_2 flow battery module composed of four stacks. It is seen that the plumbing layout for electrolyte circulation can have a significant effect on the flow rate distribution of electrolyte through the stacks.

443837 Blowdown Dynamic Modeling and Analysis of Pressure and Temperature

Abstract:

Vessel blowdown is one of the safety operation procedures that discharges flammable gas or liquid to flare system during emergency situation such as fire. When blowdown occurs, the liquefaction and freezing effect cause a pressurized vessel to be brittle, exposing it to possible hazard situation. Also the blowdown systems are recommended to be accordance with API(American Petroleum Institute) 521 code. API 521 requires pressure in vessel should be half of the initial pressure or be lower than 100 psig in 15 minutes after commencing blowdown in fire case. Therefore, accurate estimation over the change of temperature and pressure of the vessel during blowdown period is crucial for the design of blowdown system. For the reason, researchers have developed a few dynamic blowdown models so far. However, the model was found to be rather inaccurate under certain conditions. Therefore a new dynamic blowdown model was developed in this work. For the accurate analysis, the model algorithm considers both vapour and liquid phase in non-equilibrium condition that the temperatures of vapour and liquid phases are different. Futhermore, heat transfer from the vessel wall to discharge gas at either laminal or turbulent flow in the vessel was considered in the model to increase the accuracy of blowdown estimation. For thermodynamics, both Peng-Robinson and Soave-Redlich-Kwong equations of state were applied to the model, and the results were compared each other. Finally, the simulation results were compared with Haque et al's experimental and steady-state simulation results.

443974 Observer-Based Model Predictive Control for Time-Delay Systems

Abstract:

Model predictive control is one of the most widely used advanced process control technologies in industry.[1] It has many capabilities such as dealing with input/state constrains and uncertainties. Observer-based model predictive control was developed to estimate states information when the states are not measurable directly. Many chemical processes have time delays in either states or measurements, such as polymer production and some bioprocesses. [2] In this presentation, we are proposing an unknown input observer based model predictive control algorithm for time-delay systems. Unknown input observer is an observer design to eliminate certain types of unknown inputs in the system. Stability of unknown input observer-based model predictive control without time delay was provided by our previous study. Terminal cost and terminal region are designed for a class of non-linear time delay systems. The close-loop stability of the controller is provided. [3][4]

An example of bioreactor is provided which has uncertain time delays in measurements but with a known upper bound. Two measurements are provided which are online biomass and offline ethanol measurements. Model-plant mismatch is considered as unknown inputs in the system. State estimation and control are calculated using the observer and an optimization algorithm.

Reference:

[1] Marcus Reble et al., Model predictive control of constrained non-linear time-delay systems. IMA Journal of Mathematical Control and Information, 2010, 1-19

[2] Seung Cheol Jeong and PooGyeon Park, Constrained MPC algorithm for uncertainty time-varying systems with state delay, IEEE TRANSACTIONS ON AUTOMATIC CONTROL, 50, 2005, 257-262

[3] Mayuresh Kothare et al., Robust constrained model predictive control using linear matrix inequalities. Automatica, 32, 1996, 1361-1379

[4] Wook Hyun Kwon et al., A simple receding horizon control for state delayed systems and its stability criterion, Journal of Process Control, 13, 2003, 539-551

444020 Thermodynamic Modeling of Aqueous Lithium Sulfate and Its Solubility in the Presence of Na+, K+ and Mg2+ Sulfates

Abstract:

This work is an extension of the previous works done by our group towards development of a rigorous and comprehensive thermodynamic model for high salinity produced water that is essential for mass and heat balance calculations and simulation and design of novel treatment processes. Specifically this work presents a thermodynamic model that accurately represents all thermodynamic properties of aqueous lithium sulfate solution including the solubility of Li2SO4 in water and also in the presence of Na+–K+–Mg2+ sulfates. By explicit consideration of solution chemistry of each subsystem, symmetric electrolyte NRTL model is used to correlate available experimental data of all pertinent thermodynamic properties including activity coefficient, osmotic coefficient, excess enthalpy, and liquid heat capacity over the entire concentration range. This facilitates quantification of eNRTL binary molecule–molecule, molecule–electrolyte, and electrolyte–electrolyte interaction parameters. Also, solubility data are used to regress the thermodynamic constants required for modeling the liquid–solid equilibrium of lithium sulfate in water and in the presence of Na+–K+–Mg2+ sulfates. Displaying a significant prediction ability, the model accurately represents the thermodynamic behavior of the system from infinite dilution up to saturation and temperatures from 298.15 K to 573.15 K.

444044 Organic Solar Cells

Abstract:

This paper describes highlights the need of next-generation "organic solar cells". There is a continuing need for organic solar cells and arrays with very high specific power densities (1000-5000 kW/kg) for generating power in a new generation of ultra light space payloads example in NASA. An emerging technology with promise to meet these ambitious goals are organic solar cells based on very thin films of discrete layers. Theoretical efficiencies for multi junction solar cells are more ~70% as compared to ~30% for single junction devices. There is continuous development going on high mobility, band gap engineered organic donors and acceptors, as well as development of three junction devices and scaled up devices on lightweight flexible polymer substrates. The main interest in this paper is organic semi-conducting crystal: Pentacene, which has photo-conducting states. Organic solar cells manufactured today have low efficiency other than their stability issues. The efficiency of organic solar cells can be improved by either designing polymer materials that absorbs more of solar spectrum or by minimizing the charge collection efficiency by controlling the morphology of polymer films. Once fully developed, organic solar cells would revolutionary power the industry due to their flexibility and minimal weight

444055 A Refined Non-Random Two-Liquid Segment Activity Coefficient Model for Solubility Modeling

Abstract:

Accurate correlation and robust prediction for complex molecule solubility or solid-liquid equilibrium (SLE) in solvents and solvent mixtures have drawn much interest recently [Tanveer et al., *Chem. Eng. Prog.* **2014**, 110 (9), 37-47]. Solubility models such as the Non-Random Two Liquid Segment Activity Coefficient (NRTL-SAC) model [Chen and Song, *Ind. Eng. Chem. Res.* **2004**, 43, 8354-8362] are now routinely used in the industries. However, an important unfilled need is to model the solubility in oligomers/polymers/surfactants such as asphaltene nanoaggregates in petroleum fluids and drug molecules in lipid-based or solid dispersion formulation [Tung, *Org. Proc. Res. Dev.* **2013**, 17, 445-454]. This paper presents a refinement to the very successful NRTL-SAC model to better account for the contribution due to size differences of molecules. Specifically, the Flory-Huggins combinatorial entropy expression for conceptual segments in the model is substituted with the Staverman-Guggenheim expression for real molecules. We present the rationale for the refined model, the updated conceptual segment numbers for common solvents, and the newly identified conceptual segment numbers for common drug excipients. The model performance is illustrated by prediction of drug molecule solubility in various lipid solvents and solvent mixtures, compared to experimental data and previous work by other models. The refined NRTL-SAC model shows enhanced capability to correlate and predict drug molecule solubility in common solvents and excipients used in drug formulation.

444057 Explaining the Paradox of Asphaltene Precipitation with n-Alkanes

Abstract:

Explaining the paradox of asphaltene precipitation with n-alkanes

Md Rashedul Islam and Chau-Chyun Chen
Department of Chemical Engineering, Texas Tech University

Abstract
Paraffins, specifically alkanes, promotes asphaltene precipitation when they are blended with asphaltene-containing petroleum fluids. The alkane/oil volume ratio required for onset of asphaltene precipitation increases with the carbon number and yields a maxima. Experimental study indicates that the maxima lays at C7[1] or between C8 -- C10 [2]. However, precipitation yield decreases asymptotically with increase of the carbon number even if excessively high alkane/oil volume ratio is used [3]. For example, n-pentane be a better solvent for asphaltene than n-hexane in regard of both onset of precipitation and precipitation yield. However, high carbon number alkane beyond the maxima, n-hexadecane for instance, can be a better solvent at onset of precipitation point and a poorer solvent in regard of precipitation yield than n-pentadecane. A thermodynamic explanation for this paradox is necessarily important to find the compatibility of bitumen or crude oil with paraffinic hydrocarbons. In this work, this paradox is explained by the aggregation thermodynamics of asphaltene precipitation [4]. Both the enthalpy and entropy of mixing are investigated to explain the effect of paraffin carbon number on asphaltene precipitation. Athabasca and cold lake bitumen samples are studied.

References

[1] G. Hotier and M. Robin, "Action de divers diluants sur les produits pÃ©troliers lourds: mesure, interprÃ©tation et prÃ©vision de la floculation des asphalténes," Oil & Gas Science and Technology, vol. 38, pp. 101-120, 1983.
[2] I. A. Wiehe, H. W. Yarranton, K. Akbarzadeh, P. M. Rahimi, and A. Teclemariam, "The paradox of asphaltene precipitation with normal paraffins," Energy & fuels, vol. 19, pp. 1261-1267, 2005.
[3] D. L. Mitchell and J. G. Speight, "The solubility of asphaltenes in hydrocarbon solvents," Fuel, vol. 52, pp. 149-152, 1973.
[4] M. Wang, Y. Hao, M. R. Islam, and C. C. Chen, "Aggregation Thermodynamics for Asphaltene Precipitation," AIChE Journal, Submitted.

444086 Optimization of Layered Graphene Oxide Nanostructures through Controlling the Membrane Deposition Rate Towards Size-Sieving Separation

Abstract:

Single-layer graphene oxide (GO) flakes were coated on anodized aluminum oxide (AAO) substrate through a simple vacuum filtration process, the effect of the deposition rate on the nanostructures of obtained membrane was investigated though pure water flux and organic vapor diffusion, which indicated GO membrane assembled with slow deposition rate might have more ordered lamellar structure and, correspondingly, have smaller effective pore size. In contrast, under fast deposition rate, GO flakes assembled in a more random way. Therefore, through controlling the membrane deposition rate, the nanostructure of layered GO membrane could be optimized for size dependent separation. We found after thermal reduction, 100 nm GO membrane (prepared @ 0.02 nm/s) has effective pore size smaller than 6.1 Å, which exhibit great potential in n-hexane/2,2-dimethybutane (DMB) separation.

444142 Robust Design of a Radiant Wall Burner for a High Temperature Environment Utilizing Computational Fluid Dynamics Simulations

Abstract:

Radiant Wall Burners are often utilized in Ethylene Cracking Furnaces. The conditions and fuel gases for a specific unit can vary considerably within the unit and across similar units. With burner air preheated to high temperatures, premixing the fuel can result in dangerous and destructive flashback, while complete nozzle-mixing of the fuel can result in burner instability at startup or turn-down conditions. These challenges can be solved through the generation of a robust radiant wall burner design, which is facilitated by executing Computational Fluid Dynamics simulations of many different geometric configurations.

444161 Novel Microwave Synthesis Carbon Dots from PINE Bark and Its Application for Heavy Metal Sensing

Monday, April 11, 2016
Exhibit Hall E (George R. Brown)

Saheed Sanni, Chemistry, Vaal University of Technology, Vanderbijlpark, South Africa, Puthalapattu ReddyPrasad, Chemistry, Vaal University of Technology, Vanderbiljpark, South Africa and Augustine E. Ofomaja, Chemistry, Vaal university of Technology, Vanderbiljpark, WV, South Africa

NOVEL MICROWAVE SYNTHESIS CARBON DOTS FROM PINE BARK AND ITS APPLICATION FOR HEAVY METAL SENSING

Sanni S.O., ReddyPrasad P., Ofomaja A.E.

Biosorption and Waste water Treatment Research Laboratory, *Department of Chemistry, Faculty of Applied and Computer Sciences, Vaal University of Technology, P. Bag X021, Vanderbijlpark-1900, South Africa.*

INTRODUCTION

Microwave pyrolysis for the production of carbon has the advantages of being cheap, produces high yield of carbon and is absolutely a green synthetic method [1]. In recent times, the use of microwave pyrolysis have been extended to the preparation of carbon dots. Carbon nanodots (C-dots) falls under the class of carbon nanoparticles with size below 10 nm[2]. C-dots are carbon materials which processes high water solubility and strong fluorescence depending on their size, edge, shape, surface ligands, and defects. Since C-dots preparation can be achieved from difference carbonaceous sources and by varying a range of parameters during their synthesis, the properties of the final product may vary depending on these factors. C-dots have sp^2 characteristics and possess several functional groups on their surfaces that makes them highly hydrophilic and easily functionalized with various organic, polymeric, inorganic, or biological spices[3]. Several synthesis methods that have been explored includes electrochemical synthesis, combustion/thermal/hydrothermal/acidic oxidation, supported synthesis, microwave/ultrasonic, arc discharge, laser ablation/passivation, and plasma treatment[4]. C-dots processes excellent such as water dispersibility, chemical stability and photostability, ease of modification, excitation-dependent multicolor emission, low toxicity, and good cell permeability, great efforts have been paid to their potential applications in biosensing, bioimaging, drug delivery, and other biological related aspects.

METHODS

The carbon dots (C-dots) with green luminescence was synthesized from microwave assisted pyrolysis of pine bark waste our carbon source at 800, 1000 1200 W for 1 Hr. The dots produced were characterized using FTIR, TGA, XRD, UV-Vis and Raman spectroscopy. The dots were applied for heavy metal sensing.

RESULTS AND DISCUSSION

The functional groups present on the C-dots (**Fig.1**) includes -OH stretching band mode at 3413 cm^{-1}, C-H stretching at 2934 and 2845 cm^{-1}, stretching peaks of C=O, C=C at 1731, 1604 and 1437 cm^{-1}. The XRD pattern (**Fig 2a &b**) of produced C-dots depicts sharp peak at 20.9^{o} and 26.8^{o}, which is highly crystalline cores and possess amorphous nature of C-dots. This peculiar peak is ascribed to 002 of the graphitic carbon, which in line with inter-planar spacing of 3.77 Å with higher spacing than bulk graphite at 3.34 Å. The thermogravimetric and differential thermal analysis display (**Fig 3**) that the C-dots possess thermal stability more than 100 ^{o}C and mass losses occurs afterwards. The thermal analysis depicts 3 endothermic peaks at 331, 542 and 735 ^{o}C, which are ascribed to oxygenated groups such as carboxyl, hydroxyl, carbonyl and phenols. The optical properties of the synthesize C-dots were explored using UV-vis absorption and photoluminescence (PL) spectroscopy. The UV-vis spectra display (**Fig 4**)intense absorption peak in the range of 260 to 290 nm, which is ascribed to π-π^* transition of the aromatic sp2 domain and also presence of multiple polyaromatic chromophores. A shoulder peak at 330 nm which is ascribed to n-π^* transitions of C=O and structural defects of the C-dots. PL spectrum of the C-dots (**Fig 5a**)exhibits an intense maximum emission peak at 432 nm upon excitation at 330 nm, which also shows characteristic yellowish green color under sunlight and intense blue coloration UV lamp (365 nm). The PL properties of as prepared C-dots was varied at (**Fig5b**) different excitation wavelength from 300 to 405 nm. There was red-shifting in emission peak as the excitation peak was increased from 300 to 405 nm and also simultaneously decrease in fluorescent intensity, which is ascribed to quantum effect and surface defects. This behavior of the prepared C-dots is dependent on excitation-fluorescence, which is accordance with other report [13]. The quantum yield of synthesized C-dots was measured to be 17% at maximum excitation at 330 nm

Fig.6 shows the Raman spectrum of the PB-GQDs, which has been usually used to confirm the quality of the prepared PB-GQDs. There is two major Raman peaks at 1370 cm^{-1} (D band), 1580 cm-1 (G band). The peak of G band corresponds to the Sp^2 bonded carbon atoms in a two-dimensional (2D) hexagonal lattice and the peak of G band is associated with the vibrations of carbon atoms with dangling bonds in the termination plane of tangled graphite or glassy carbon. The D/G ratio band ratio is ID/IG: 0.68 is confirm the formation of graphene like structure from synthesized pine bark.

Fluorescence detection of Cu^{2+} ion

The detection of Cu^{2+} was performed at room temperature in phosphate buffer (PBS) solution pH 4.3. 1.0 mL of PB-CQDs solution was added into 0.5 mL of PBS buffer, followed by the addition of 1.0 mL different concentrations of Cu^{2+}. The fluorescence emission spectra were recorded after reaction for 5 min at room temperature (~28 °C).

Selectivity and sensitivity studies

Moreover sensitivity and selectivity is additional significant parameter to assess the performance of the sensing system. Consequently, we examined the fluorescence intensity changes in the presence of representative metal ions such as including Cu^{2+}, Hg^{2+}, Ca^{2+}, Al^{3+}, Ba^{2+}, Mn^{2+}, Ni^{2+}, Pb^{2+}, Co^{2+}, Zn^{2+}, Fe^{3+}, $Cs^{2+}$and Mg^{2+}, of each at a concentration of 15.0 µg mL^{-1} were reacted with a PB-CQDs solution. Fig. 7 A displays the relative change in PL intensity of PB-CQDs in the presence of various metal ions. Fe^{2+}, Mn^{2+} and Hg^{2+} ions cause the slight PL changes (defined as the relative change of PL intensity in 60–80% as compared to blank) which can be attributed to the nonspecific interactions between the carboxylic and/or amine groups and the metal ions. Among these metal ions, Cu^{2+} displays the strongest PL quenching effect on PB-CQDs, assigning to the special coordination between the Cu^{2+} ions and the phenolic hydroxyl and/or amine groups of PB-CQDs which has been widely used for the detection of Cu^{2+} ions or colored reactions in traditional organic chemistry. Such a specific fluorescence quenching effect may originate from the strong interactions between Cu^{2+} ions and the surface groups of PB-CQDs which transfer the photoelectrons from PB-CQDs to Cu^{2+} ions. These results clearly demonstrate that the PB-CQDs-based Cu^{2+} sensor is highly selective to Cu^{2+} over the other metal ions. Fig.8. Fluorescence responses of CCDs to the addition of different metal ions under acetate buffer (pH 4.3). The concentration of each metal ion is 15 µg $mL^{-1.}$ F_0 and F correspond to the fluorescence intensities of PB-CQDs at emission wavelength of 440 nm excited at 320 nm in the absence and presence of metal ions, respectively.

As shown in Fig. 3, the fluorescence intensity of PB-CQDs gradually decreased with the Cu^{2+} concentration. The fluorescence intensity displayed liner responses with the Cu^{2+} concentrations, with two linear ranges of 0.0□14.0 µg mL^{-1} and 15.0□25.0 µg mL^{-1} (Fig.3B). The linear equations are $F/F0$ =0.9827□0.8485C_{Cu2+} (R = 0.993) and $F/F0$ = 0.7565–0.0033 C_{u2+} (R = 0.996), respectively.

REFERENCES

[1] W. B. Lu, X. Y. Qin, S. Liu, G. H. Chang, Y. W. Zhang, Y. L. Luo, A. M. Asiri, A. O. Al-Youbi, X. P. Sun, Economical, green synthesis of fluorescent carbon nanoparticles and their use as probes for sensitive and selective detection of mercury(II) ions, Anal. Chem. **2012**, 84, 5351-5357.

[2] J. Jiang, Y. He, S. Li, H. Cui, Amino acids as the source for producing carbon nanodots: microwave assisted one-step synthesis, intrinsic photoluminescence property and intense chemiluminescence enhancement. Chem Commun **2012**, 48, 9634–6.

[3] Y. Wang, A. Hu, Carbon quantum dots: synthesis, properties and applications, J. Mater. Chem. C **2014,** 2, 6921–6939.

[4] C. Liu, P. Zhang, X. Zhai, F. Tian, W. Li, J. Yang, et al., Nano-carrier for gene delivery and bioimaging based on carbon dots with PEI-passivation enhanced fluorescence, Biomater. **2012,** 33, 3604-13.

Figures:

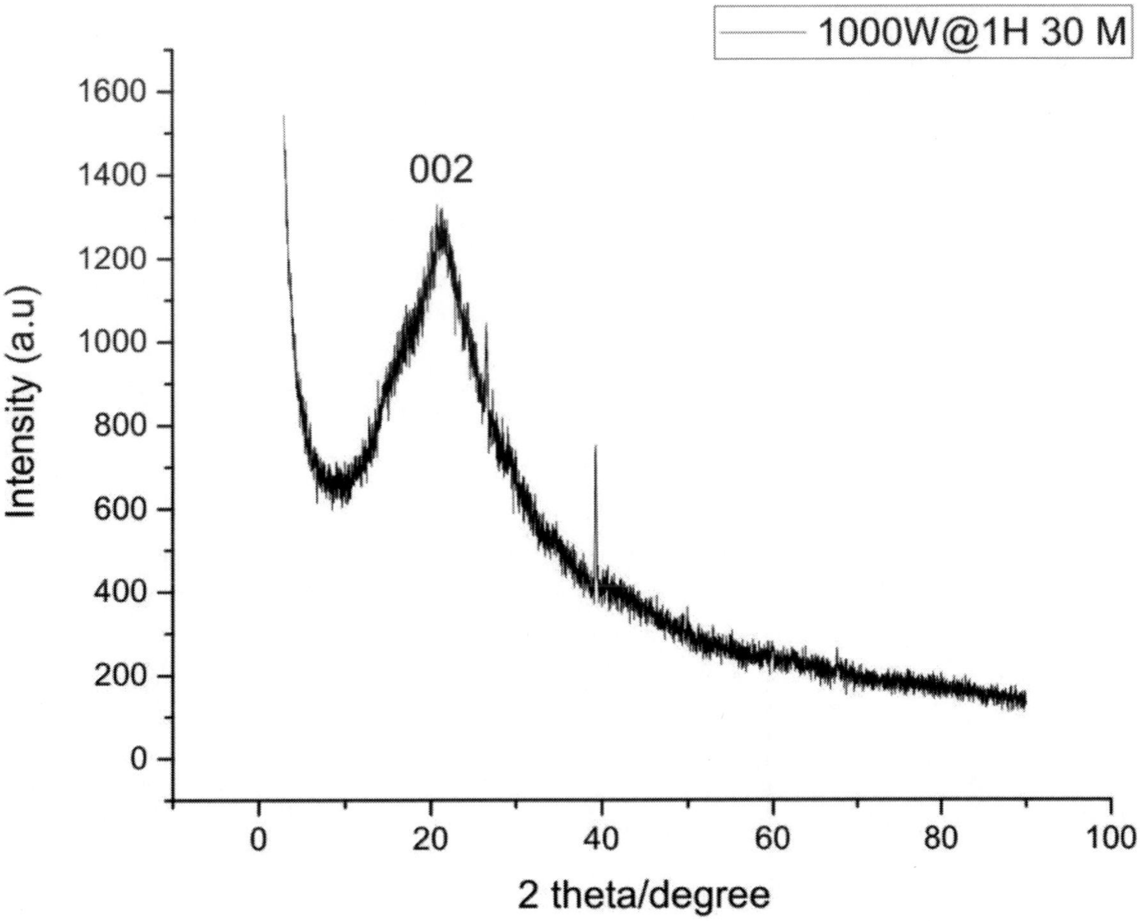
1000W@1H 30 M
002
Intensity (a.u)
2 theta/degree
1600
1400
1200
1000
800
600
400
200
0
0
20
40
60
80
100

Fig 1. FTIR spectra of as-prepared C-dots from Pine Bark

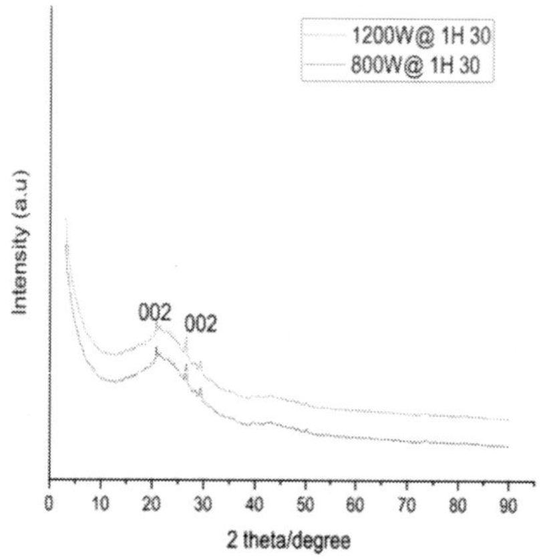

(a)

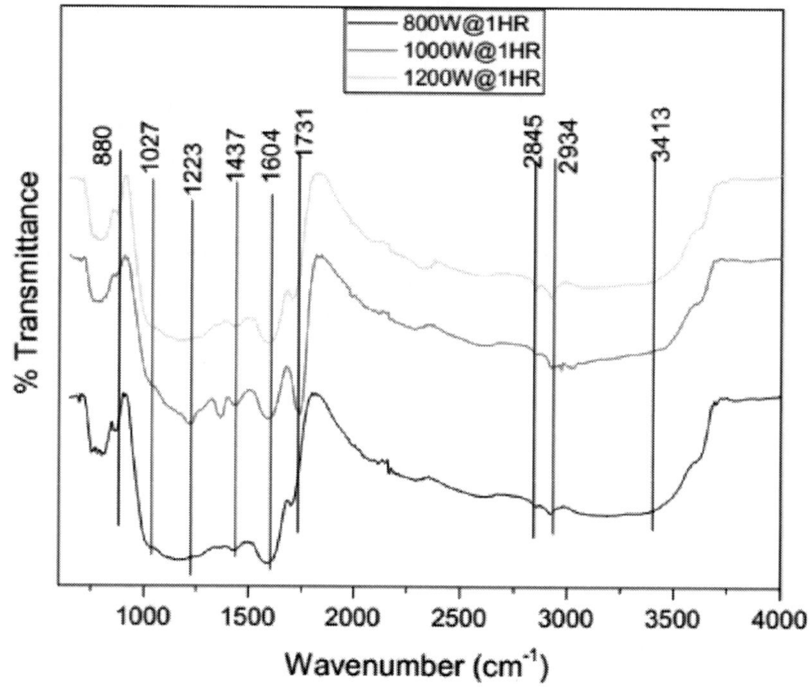

(b)

Fig. 2a & b. XRD pattern of C-dots.

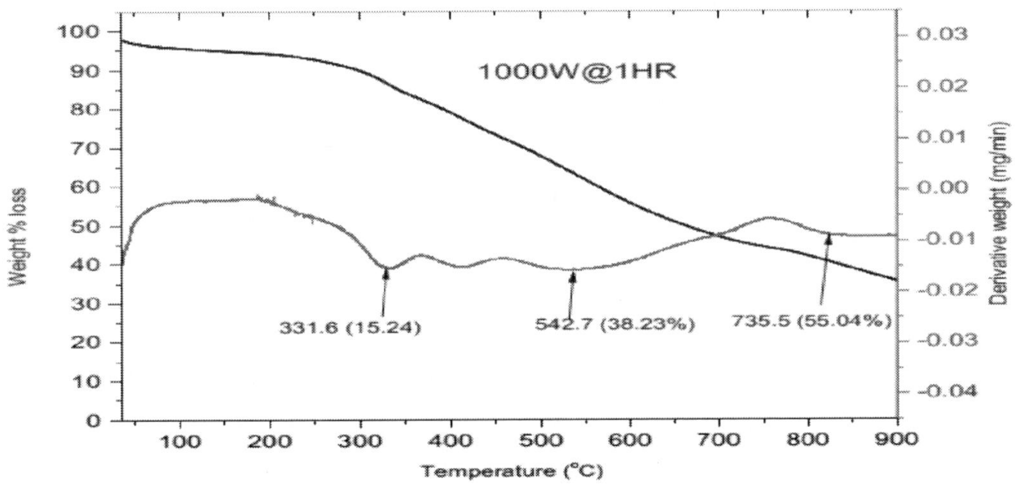

Fig 3. Thermal curves for C-dots under N_2 atmosphere.

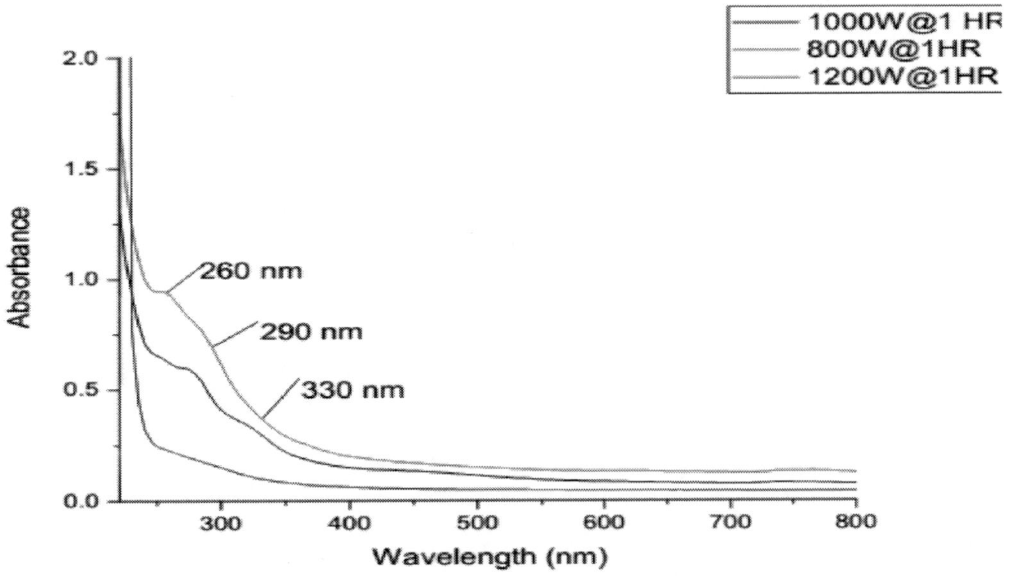

Fig 4. UV-vis absorption of C-dots.

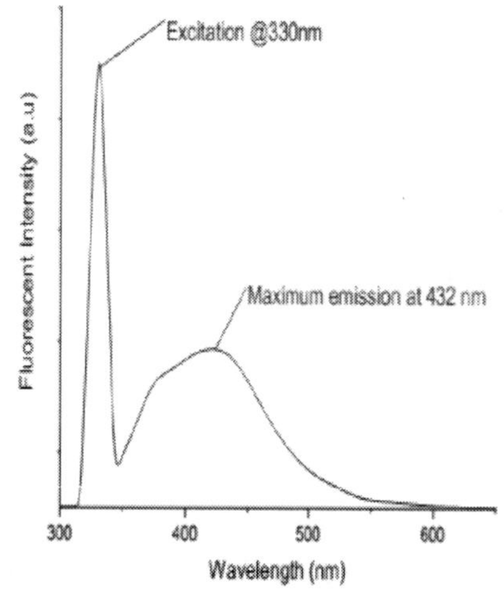

(a)

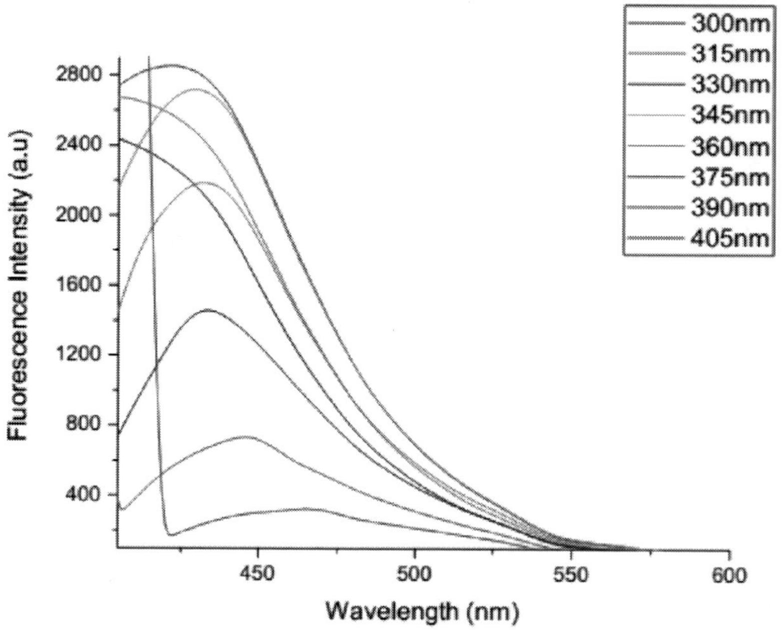

(b)

Fig 5a. PL excitation and emission wavelength (b) PL emission spectra at different excitation wavelength from 300 to 405 nm in 15 nm increments.

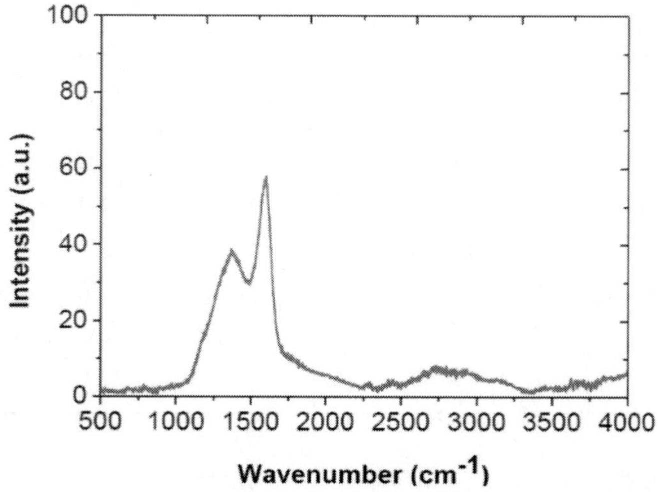

Fig. 6. Raman Spectra of PB-CQDs

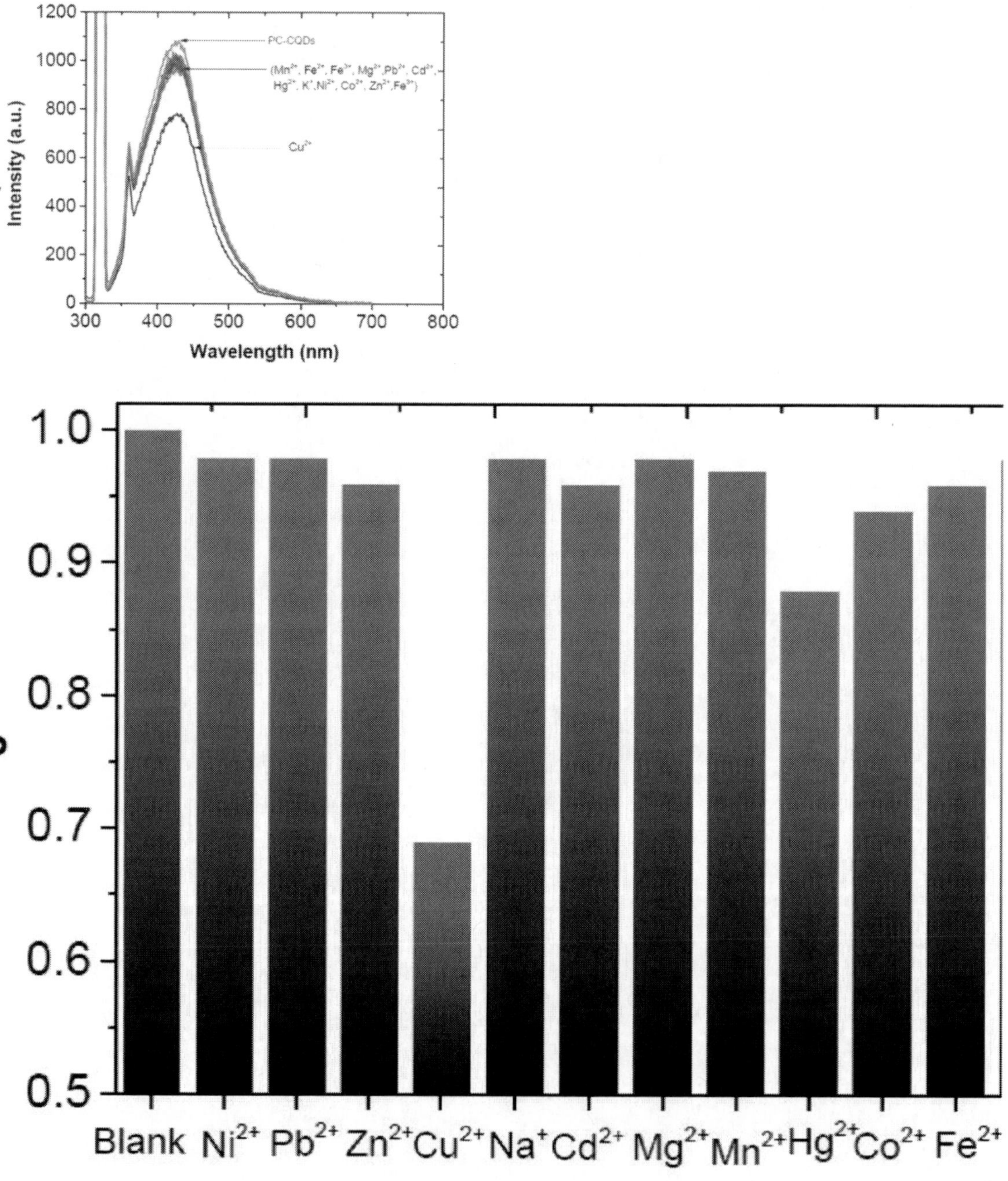

Fig.7. Fluorescence responses of CCDs to the addition of different metal ions under acetate buffer (pH 4.3).

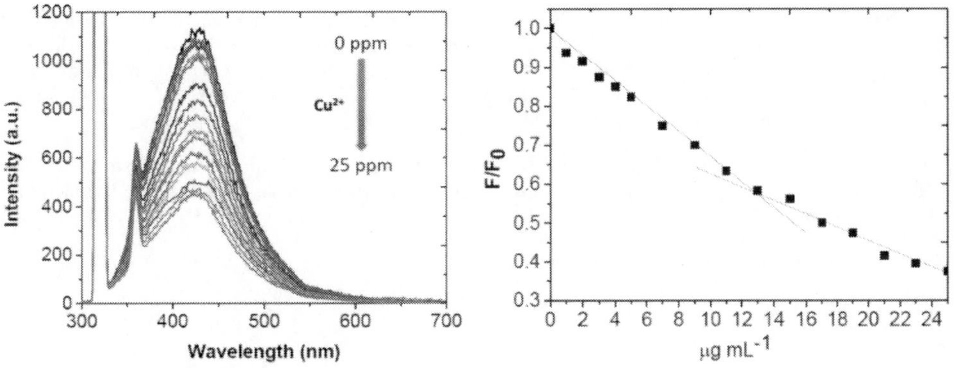

Figure 8. Fluorescence titrations of the y-CDs with Cu^{2+} from 0 to 25 μg mL^{-1}. (Influence of Cu^{2+} concentration on the fluorescence intensity of PB-CQDs (top to bottom: 0, 1, 2, 3, 4, 5, 7, 9, 11, 13, 15, 17, 19, 21, 23 and 25.0 μg mL^{-1})

444269 Functionalization of OH-Terminated Nanoparticles By Cyclic Azasilanes

Abstract:

Colloidal suspensions of functionalized nanoparticles have been suggested for enhanced oil recovery, formation damage remediation, flow assurance, and water treatment operations. Surface modified nanoparticles can also be used as tracers in reservoir studies, active media in thermal management fluids, and fillers in high performance elastomers. Conventional functionalization of nanoparticles with alkoxysilanes and chlorosilanes is limited by needs for catalytic amount of water for the reaction to proceed, requirement of multiple hours to fully saturate the surface, and release of by-products upon reaction completion. Monolayer deposition with a high density of functional groups represents a particularly difficult challenge.

Our ongoing studies indicate that application of cyclic azasilanes that have a strained ring structure with adjacent Si and N centers provides highly efficient functionalization of hydroxylated silica, alumina, magnetite, and diamond nanoparticles which is catalyst-free and by-product-free. The data on synthesis and characterization of functionalized nanoparticles as well as possible reaction mechanism of functionalization will be discussed.

444093 New Development on Hoist Scheduling with the Relaxation of Cyclic Precondition

Abstract:

Multi-Stage Material Handling (MSMH) processes are designed to manufacture materials requiring a series of treatments and in industry, they are broadly used due to their capability to handle large work orders and compatibility with materials with different recipes. A typical example is electroplating process. A hoist handles the transportation of each job among units (stages) according to its processing recipe. The hoist movement plan is predetermined through optimization and this plan is repeated to perform continuity. Obviously, the ultimate goal is to maximize the productivity within each movement plan, which is equivalent to minimize the cycle time to finish one work order. The optimization process is generally called Cyclic Hoist Scheduling (CHS), since every movement plan is a cycle due to its repeatability. This precondition in term of hoist operation and unit usage is general for most cases. However, it could block the really optimal results.

In this study, this precondition has been relaxed for multi-capacity units, thus the unit usage can be more flexible. Thus, the real optimality can be found through CHS. To deal with the cycle connection issue from the relaxation of the precondition, a new module Cycle Connection Identification (CCI) is proposed. Obtained hoist schedule will be checked. Once a schedule with cycle connection problems is identified, a new cycle will be scheduled forcing the smooth connection with the first cycle. The objective of scheduling a new cycle is to match the beginning of the first cycle and the ending of the second cycle. If the two-cycle group is still with cycle connection issue, more cycles will be scheduled till the cycle group is free from cycle connection problems. Each cycle is called a sub-cycle and the constructed cycle group without cycle connection problems is called a full cycle. Note that in each sub-cycle, the work order and all transportation operations are completed.

In this paper, hoist scheduling models with/without such a precondition have been developed and proposed as mixed integer linear programming. The performance of the two models is tested through case studies and the difference is compared and demonstrated in detail.

444359 Development of Novel Emulsified Acid Using Waste Oil

Abstract:

In carbonate reservoirs, hydrochloric acid (HCl) is commonly used for stimulation purposes. The main disadvantage for using HCl is its high and rapid reactivity allowing for less control on acid reaction with formation. In addition, HCl may react with the tubing causing severe corrosion and damages. Several slow-reacting acid systems, though HCl-based, have been developed to retard the HCl reaction and control its reactivity to allow deeper penetration in carbonate reservoirs before it breaks out and reacts with formation rock. However, the selection of the hydrocarbon phase of emulsified acid is critical to the creation and stability of the emulsified acid system. Diesel oil, crude oil and solvents such as xylene have been reported to be successful in the field.

In this work, alternative and less expensive source of oil is developed to replace the current practices. Emulsified acid is a water-in-oil emulsion where oil is the continuous phase and acid is the dispersed one. Industrial and filtered waste oil was used to prepare the emulsified acid of this work. Extensive work has been carried out to study the chemical composition of the waste oil, namely SARA and to select the proper emulsifier. Emulsions were prepared using 15% HCl with emulsifier concentration ranging from 0.5 to 2.0 vol% and 70:30 of acid-to-oil ratio. Droplet size distributions were measured and used to benchmark with other emulsions. Emphasis was given to the stability of the emulsion especially at high temperatures. In order to apprehend the effect of corrosion inhibitor on the emulsion stability, its concentration was varied (up to 0.3 vol%). Later, rheological analyses of the waste oil emulsion were compared to the commonly used diesel-based emulsified acid.

Results showed the potential use of waste oil to prepare high temperature emulsified acid (up to 100°C). The addition of corrosion inhibitor to the emulsified acid caused a negative effect on stability resulting in a less stable emulsion. On the other hand, addition of cationic emulsifier helped to formulate a more stable emulsion. Apart from that, the optimum concentration of the emulsifier was at 2.0 vol%. The rheological data confirms the stability of the new emulsified formulation at higher temperatures. This paper summarizes the findings of using waste oil emulsion and recommends it for field applications. The newly proposed emulsified acid introduces waste oil as a replacement of diesel.

The novelty of using waste oil will help to significantly reduce the expenses of formulating emulsified acid. Since waste oil is basically waste it has a good impact on the environment, that being said, instead of throwing it, the industry will be able to recycle it.

444364 Molecular Simulation of Controlling Ions and Their Role in Low Salinity IOR for Carbonate Reservoirs

Abstract:

Carbonate geology represents around 50% of hydrocarbon-bearing reservoirs in the world and more than 70% in some areas such as the Middle East. Water flooding has not been very efficient in carbonate reservoirs due mainly to the low water wetness. Modifying the ionic composition and salinity of the injected seawater demonstrated that the oil recovery from these carbonate reservoirs could be improved. Although the underlying mechanisms are still a subject of debate, the improved oil recovery has been mostly attributed to the increased water-wetness of the rock through the interactions of specific ions like Ca^{2+}, Mg^{2+}, SO_4^{2-}, and Cl^- , which are present in seawater with the rock surface under favorable conditions

The study of hydrated gas-phase metal ions provides a link between intrinsic chemistry of the isolated ion and its chemistry in solution. The structures and binding enthalpies of a variety of gas-phase and solution mono- and divalent ion hydrates containing up to 12 water molecules have been studied computationally. A DFT and molecular dynamics simulations were employed to study hydrated structures of the form $[\mathbf{M}(H_2O)_n]^{2+}.mH_2O$ for calcium and magnesium and $[\mathbf{M}(H_2O)_n]^{+}.mH_2O$ for sodium, in which there are n water molecules in the first coordination shell and m water molecules in the second coordination shell. The hydration, dehydration, and interactions of metal ions with Cl^{-1} and SO^{-2}_4 have been studied to better understand the mechanism of low salinity IOR. Density functional (DFT) theory using the B3LYP functional, and 6-311++g** basis set has been used. The charge of the metal ion decreases monotonically as the number of water molecules increases in the complex. Extensive calculations of the energetic contribution of individual structural characteristics as well as the energetic trends in binding energy are developed, which will lead to better understanding of how low salinity water IOR works.

Keywords: Hydration, carbonate reservoir, DFT, binding energy, Low Salinity

444278 Insight into the Mechanism of Toxic Chromium (VI) from Wastewater Using Magnetite Coated PINE Powder

Abstract:

INSIGHT INTO THE MECHANISM OF TOXIC CHROMIUM (VI) FROM WASTEWATER USING MAGNETITE COATED PINE POWDER

Pholosi A., Ofomaja A., Naidoo E.B.

Biosorption and Wastewater Treatment Research Laboratory, Department of Chemistry,

Faculty of Applied and Computer Sciences, Vaal University of Technology, P. Bag X021, Vanderbiljpark, 1900 South Africa.

INTRODUCTION

A knowledge of the interaction between pollutant molecules or ions and active sites on an adsorbent surface is an important phenomenon in many industrial processes such as catalysis, adsorption and other separation techniques. The use of lignocellulosic materials for Cr(VI) adsorption has gained interest over the years **(Miretzky and Cirelli, 2010)**. The adsorption properties of lignocellulosic materials has been attributed to acidic functional groups on their surfaces which can take part in ion-exchange, complexation, chelation and hydrogen bonding with pollutant species in solution **(Vithanage et al., 2015)**. Two mechanisms for lignocellulosic adsorption of Cr(VI) have been proposed; (i) an electrostatic mechanism in which at low pH, the protonated acidic functional groups on the biomaterial carrying a positive charge attracts the negatively charged Cr(VI) species **(Demirbas, 2005)** and (ii) the adsorption-coupled reduction mechanism in which Cr(VI) is reduced to Cr(III) either in solution (Direct reduction) or after being adsorbed (Indirect reduction) by electron-donor groups of the biomaterial that have lower reduction potential values than that of Cr(VI) **(Gao et al., 2008)**. The Cr(III) ions formed in solution may then form complexes with Cr(III)-binding groups on the biomaterial surface or released into bulk solution by repulsion between Cr(III) and other positively charged groups on the biomaterial **(Miretzky and Cirelli, 2010)**.

METHODS

Sodium hydroxide treated pine coated with magnetite (NTP-NC) was prepared by co-precipitation method. The effect of solution pH was performed in a batch system in which 0.5 g of the adsorbent was contacted in six 250 cm^3 beakers containing 75 cm^3 of 75 mg/dm^3 Cr(VI) solutions set at pH 1, 2, 4, 6, 8 and 10 at 26 °C and agitated at 200 rpm for 2 h and the clear solution analysed for Cr(V/I) left using UV-Vis spectrophotometry at 460 nm

RESULTS

The results of the effect of initial solution pH on the adsorption of Cr(VI) onto NTP-NC is shown in Fig. 1. To confirm the operating mechanism, the change in $[H^+]$ concentration at each initial pH, the oxidation/reduction potential before and after adsorption and the amounts of Total Cr, Cr(VI)and Cr(III) present in solution at each pH was monitored for NTP-NC and the results displayed in Fig1 . The change in $[H^+]$ concentration ($\Delta[H^+] = [H^+]_{final} - [H^+]_{initial}$) at the end of the adsorption process for each initial solution pH was recorded and plotted against the initial solution pH. The results showed that the change in $[H^+]$ concentrations at the end of the adsorption process where higher for lower initial solution pH's (where lower amounts of Cr was found in solution) than for higher initial solution pH's (where higher amounts of Cr was found in solution) and the change in $[H^+]$ concentration became small and almost constant after initial solution pH 4. This is consistent with the assumption that the removal of Cr from solution is associated with $[H^+]$ ion consumption and that as the concentration of $[H^+]$ ions reduces in solution, the Cr removal reduces. Cr(VI) ion in acidic solution has a positive redox potential value (above 1.3 V at standard conditions) and in the presence of an electron donor becomes unstable and can be reduced to Cr(III) **(Park et al., 2008)**. The results show that the initial ORP values (before adsorption) for the Cr solutions at pH's of 1, 2, 4, 6, 8, and 10 were 339, 280, 165, 55, -53 and -172 mV for NTP-NC respectively. The positive values of ORP at lower solution pH's indicates higher possibility for reducing Cr(VI), while the lower and negative ORP values indicates lower possibility for reducing Cr(VI). On addition of the adsorbents into the solutions, a change in ORP were observed . The values of ORP at the end of the adsorption process became 330, 268, 30, 9 -29 and -61 for NTP-NC respectively. A reduction in ORP at the end of the adsorption stage signifies that reduction has occurred in the system. **Yu et al. (2014)** showed the amount of Cr(VI) removed from aqueous solution in contact with zero-valent iron (nZVI) could be related to the change in ORP and pH of the solution. The authors observed a decrease in ORP from 400 to 100 mV and a corresponding 10 unit pH increase as Cr(VI) was removed from solution.

Finally, analysis of the prepared Cr solutions at the various solution pH's showed that the concentration of Total Cr obtained from ICP and that of Cr(VI) using UV-vis measurements were very similar, suggesting that Total Cr was almost equal to Cr(VI) in solution. After contacting the various solutions with the adsorbents, the concentration of Total Cr in the pH 1 solution was observed to have reduced to 21.56 mg/dm^3 for NTP-NC, while the Cr(VI) concentrations was 10.66 mg/dm^3 for NTP-NC. With solution pH 2, the concentration Cr(III) left in solution was lower than that of solution pH 1 for both samples. The reason is that at pH 1 there is an abundance of H^+ ions in solution which accumulates on the functional groups on the adsorbent making the surfaces of the adsorbent positively charged and repelling away the positively charged Cr(III) ions from the adsorbent surfaces **(Suksabye et al., 2009)**. At solution pH 2, it is believed that there is less accumulation of H^+ ions at the adsorbent surfaces which led to reduced electrostatic repulsion and increased adsorption of Cr(III) from solution. This increase in Cr(III) adsorption is also confirmed by the reduction in Total Cr in solution between solution pH 1 and 2. The concentrations of Cr(VI) left in solution at pH 2 was higher than at pH 1 for NTP while for NTP-NC the Cr(VI) concentrations were lower at pH 2. For solution pH's higher than 2 and up to 10, the residual Cr(III) in solution becomes very low and almost constant. This is consistent with the small change in $[H^+]$ in solution which is almost constant at this pH range.

FTIR Evidence for the adsorption of Cr(III)

When Cr(III) is formed in solution via adsorption-coupled reduction mechanism it may undergo hydrolysis and/or complexation depending on the solution pH and the total Cr(III) concentration. Hydrolysis may lead to the formation of various Cr(III) species with an increase in solution pH which gives rise to a reduction in its solubility and increasing the chances of precipitation of the Cr(III) species. At solution pH 1 – 4, Cr^{3+} and $CrOH^{2+}$ are the major species which can be adsorbed by negatively charged ligands on the biomaterial surface. On the other hand, at solution pH range of 4 – 10, Cr(III) species (hydroxo complex), such as $CrOH^{2+}$, $Cr(OH)_2^+$ and $Cr(OH)_3$, are the dominant forms in the aqueous system and at high pH is readily transformed into soluble complex, $Cr(OH)_4^-$. When NTP-NC was contacted with Cr(VI) solution for 2 hr, the final solution pH was increased to pH 2.53. The FTIR spectra for NTP-NC (Fig. 2) shows the presence of characteristic Fe-O peaks of bulk Fe_3O_4 at 553.26 cm^{-1}, 564.36 cm^{-1}, 570.21 cm^{-1} and 576 21 cm^{-1}. Functional groups associated with the pine includes the broad band at 3250.99 cm^{-1} representing –OH of cellulose, 2905.17 cm^{-1} representing C-H stretching, while the peaks at 1627.69 and 1602.87 cm^{-1} shows the COO^- and C=O bonds respectively. After 2 hr contact with Cr(VI) solution, the peak at 564.34 cm^{-1} was observed to have split into several new peaks (Fig. 3b (ii)). These emerging peaks are characteristic of Fe-O bonds of maghemite (γ-Fe_3O_4), for example the characteristics peaks of maghemite was observed at 595.21, 685.65 cm^{-1} and 582.70, 634.91 cm^{-1} suggesting a change in the iron oxide phase (Fe_3O_4 to γ-Fe_2O_3) due to the conversion of Cr(VI) to Cr(III) by magnetite nanoparticles. According to **Peterson et al. (1997)**, Cr(III)(hydr)oxide produced is precipitated on the magnetite surface. In this study, α-Cr_2O_3 peak were observed at 604.63 cm^{-1} indicating that Cr(III) was formed during the adsorption process. Finally it was also observed that the carboxylate group at 1635.55 cm^{-1} was shifted to 1637.73 cm^{-1} while a new peak was observed at 1737.72 cm^{-1} suggesting that complexation between Cr(III) and Carboxylate ions also occurred.

REFERENCE

Demirbas, A. Energy Source 27 (2005) 1449–1455.

Miretzky, P., Cerelli, A.F. A review. J. Hard. Mater. 180 (2010) 1-9.

Vithanage, M., Rajapaksha, A.U., Ahmad, M., Uchimiya, M., Dou, X., Alessi, D.S., Ok, Y.S. J. Environ. Manage. 151 (2015) 443-449.

D. Park, C.K. Ahn, Y.M. Kim, Y.-S. Yun, J.M. Park, J. Hazard. Mater. 160 (2008) 422–427.

P. Suksabye, A. Nakajima, P. Thiravetyan, Y. Baba, W. Nakbanpote.. J. Hazar. Mater. 161 (2009) 1103–1108.

R.F., Yu, F.H., Chi, W.P., Cheng, J.C. Chem. Eng. J. 255 (2014) 568–576

M.L. Peterson, A.F. White, G.E. Brown, G.A. Parks Environ. Sci. Technol. 31 (1997) 1573–1576.

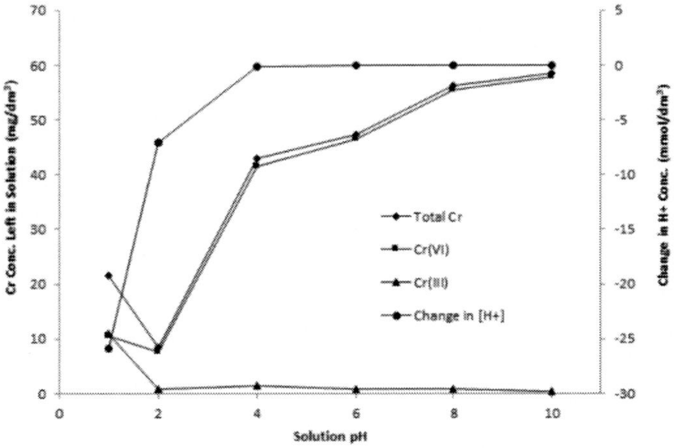

Fig. 1 Effect of solution pH on the adsorption of Cr(VI)

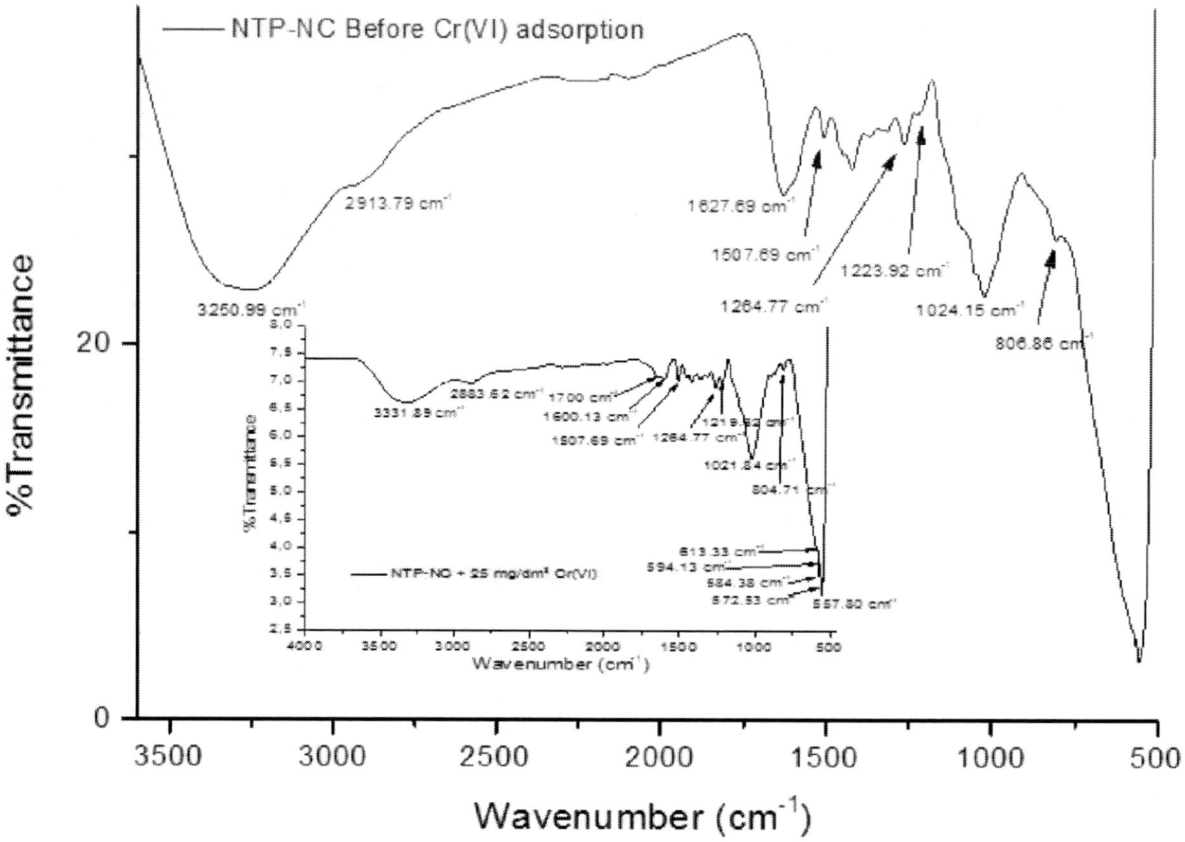

Fig, 2: FTIR of NTP-NC

444280 Copper-Phyllosilicate Core-Shell Nanocatalysts with Balanced Active Sites for Carbon-Oxygen Hydrogenolysis Reactions

Abstract:

Copper-Phyllosilicate Core-shell Nanocatalysts with Balanced Active Sites for Carbon-Oxygen Hydrogenolysis Reactions

Hairong Yue, Weimao Fang, Changmin Liu, Jiankang Ying

College of Chemical Engineering, Sichuan University, Chengdu, China

Email: hryue@scu.edu.cn

Abstract

Hydrogenolysis of carbon-oxygen (C"CO) bonds (e.g., esters, ethers, furfural, and CO_2) has emerged as a versatile synthetic tool in organic methodology, as it could produce a variety of products (e.g., chemicals, fuels, and polymers). Copper-based catalysts have been intensively explored for hydrogenation reactions as the copper sites account for the selective hydrogenation of carbon-oxygen bonds and relatively inactive for the hydrogenolysis of carbon-carbon bonds. Work on understanding the active sites of copper catalysts for hydrogenation reactions indicted that both Cu^0 and Cu^+ species were crucial to the activity of Cu-based catalysts. However, researchers were not able to establish a relationship between activity and Cu^0/Cu^+ active species since the conventional Cu-based catalysts is the deactivation by metal particle growth and unstable surface Cu^0 and Cu^+ active species in the strong reductive H_2 and oxidizing carbon-oxygen atmosphere.

We renctly present on efficient approaches for fabrication of a series of copper-phyllosilicate core-shell nanocatalysts and nanoreactors with balanced and stable Cu^0 and Cu^+ active species. We chose the hydrogenation of dimethyl oxalate, ethylene cyclicarbonate and as probe reactions for a better understanding of properties of active sites and the structure-activity relationship. The results indicated a synergy between copper and the oxide components and the balanced surface Cu^0 and Cu^+ species can greatly improve the catalytic C-O hydrogenolysis performance of the catalysts. These nanocatalysts with balanced and stable Cu^0 and Cu^+ active species, confinement effect, intrinsic high surface of Cu^0 and Cu^+ and unique tunable tubular morphology, has potential applications in the high-temperature hydrogenation reactions.

Keywords: core-shell nanocatalysts, Hydrogenation reaction; copper catalyst

References

1. Yue, H.; Zhao, Y.; Ma, X.; Gong, J.: Ethylene glycol: properties, synthesis, and applications. *Chem Soc Rev* 2012, 41, 4218-4244.

2. Gong, J.; Yue, H.; Zhao, Y.; Zhao, S.; Zhao, L.; Lv, J.; Wang, S.; Ma, X.: Synthesis of Ethanol via Syngas on Cu/SiO2 Catalysts with Balanced Cu^0-Cu^+ Sites. *J Am Chem Soc* 2012, 134, 13922-13925.

3. Zhao, S.; Yue, H.; Zhao, Y.; Wang, B.; Geng, Y.; Lv, J.; Wang, S.; Gong, J.; Ma, X.: Chemoselective synthesis of ethanol via hydrogenation of dimethyl oxalate on Cu/SiO_2: Enhanced stability with boron dopant. *J Catal* 2013, 297, 142-150.

4. Yue, H.; Ma, X.; Gong, J.: An Alternative Synthetic Approach for Efficient Catalytic Conversion of Syngas to Ethanol. *Acc Chem Res* 2014, 47(5):1483-1492

5. Yue, H.; Zhao, Y.; Zhao, L.; Lv, J.; Wang, S.; Gong, J.; Ma, X.: Hydrogenation of dimethyl oxalate to ethylene glycol on a Cu/SiO2/cordierite monolithic catalyst: Enhanced internal mass transfer and stability. *AIChE J* 2012, 58, 2798-2809.

6. Yue, H.; Zhao, Y.; Zhao, S.; Wang, B.; Ma, X.; Gong, J.: A copper-phyllosilicate core-sheath nanoreactor for carbon"Coxygen hydrogenolysis reactions. *Nat Commun* 2013, 4, 2339.

Stimuli-responsive Rubber in Oil and Gas Applications

Rostyslav Dolog, Sankaran Murugesan, and Valery Khabasheskii

Baker Hughes, 14990 Yorktown Plaza Dr, Houston, Texas, 77040, USA

ABSTRACT

A problem that exists today in the oil & gas industry is the loss of mechanical properties by reactive rubber seals due to swelling, and the limited space in downhole well applications to accommodate larger swellable packing elements.

This paper presents cost-effective, reinforcing, stimuli-responsive fillers, such as certain classes of fly ash, for use in rubber composite materials in various oil and gas applications. Apart from cost effectiveness, these fillers can also enable stimuli-responsive effects, as well as improved mechanical properties (e.g. modulus) for rubber composites when swelled in water. Fly ash (Class C) is a common component of Portland cement which was used in the literature to achieve a stimuli-responsive effect when added to Hydrogenated Nitrile Butadiene Rubber (HNBR) as filler. Exposure to water brines can cause a filler to promote an increase in pH which enables the hydrolysis of HNBR nitrile groups to carboxyl groups, thus creating an ionomer. This process is followed by neutralization of the new ionomer with metal ions, and formation of ionic crosslinks, further improving mechanical properties of the rubber.

This paper demonstrates that incorporation of fly ash as a reinforcing filler allows development of swellable seals that can retain, or even increase, their modulus and stiffness upon swelling. These characteristics allow avoiding the loss of mechanical properties by seals due to swelling and also eliminating the need for large swellable packing elements that occupy additional space downhole.

KEYWORDS: Elastomers, stimuli-responsive, fly ash, composites

INTRODUCTION

There is a growing demand for high performance elastomers in the oil & gas industry. Rubber compounds are used for various sealing applications, packer elements, stators for pumps, etc. In general, rubber formulations are comprised of elastomer matrix, curing system, antioxidants, plasticizers, along with reinforcing filler, which provides the compounds with their strength. The reinforcing filler is usually represented by carbon black and, to a lesser extent, silica. Along with other factors, filler mechanical properties and concentration generally define hardness, modulus and tensile strength of the elastomer compound. Throughout the years, many different materials have been tried as alternative fillers for rubber composites, including nanomaterials (CNTs, graphene, nanofibers), recycled rubber particles, and ash/fly ash. In the case of ordinary carbon black, the filler serves not only for reinforcement, but also as a cost reducer, because raw elastomer is often a more expensive component. Mechanical properties of the composites are affected by the aspect ratio of the filler, quality of the dispersion, and interface between blend components. The latter determines the effectiveness of load transfer from polymer matrix to the reinforcing filler. High performance compounds can be designed by incorporating well dispersed fine particles with a high aspect ratio and good compatibility with the polymer matrix.

In this context, the fly ash is a cheaper, low reinforcing filler alternative to carbon black, which will help reduce cost of the compounds for "non-demanding" sealing applications requiring large amounts of raw materials. Fly ash is produced in large quantities as a byproduct of coal burned at power plants, and can be considered as a recycled material. Fly ash costs several cents per pound, while carbon black, depending on the grade, is usually one order of magnitude more expensive. Hemmings et al[1] demonstrated that fly ash can be used as filler for polymeric composites with enhanced mechanical properties. Adoption of this technology in the oil and gas industry could change the design approach for reactive rubber compounds. Further functionalization of fly ash through a cost-efficient process could ensure compatibility with the rubber matrix, while promoting filler dispersion and ensuring an efficient load transfer. Such functionalization's include, but are not limited to, surface modification with fatty acids, dicarboxylic acids or phosphoric acid. The reasoning for surface modification is to attach functional groups with aliphatic chains to fly ash particles to make them compatible with aliphatic rubber chains, or to attach acidic groups on their surface that can interact with nitrile groups of HNBR. Initial attempts towards chemical functionalization of fly ash have been made by Seal et al for treatment of offshore oil spills.[2]

Apart from cost effectiveness, composites with fly ash can show stimuli-responsive properties. It has been shown that adding Portland cement to the HNBR matrix improves its mechanical properties in the swollen state by water uptake.[3] Exposure to water causes Portland cement to grow in volume and mass, and promotes an increase in pH which enables the hydrolysis of HNBR nitrile groups to carboxyl groups, thus creating an ionomer. This process is followed by neutralization of the new ionomer with metal ions and formation of ionic crosslinks, further improving mechanical properties of the rubber. Fly ash (Class C) is a common component of Portland cement, which contains a large variety of metal oxides, and is capable of increasing the

---

[1] Filler Comprising Fly Ash For Use In Polymer Composites. Hemmings et.al., US 6,916,863 B2, July 2005.
[2] Hydrophobic High Surface Area Zeolites Derived from Fly Ash for Oil Spill Remediation, Seal et.al., Environ. Sci. Technol., 2013, 47 (11), pp 5843–5850
[3] Stimuli-Responsive Cement-Reinforced Rubber, Musso et. al., ACS Appl Mater Interfaces. 2014 May 14;6(9):6962-8.

pH of the water solution by hydrolysis, which has a potential to achieve a stimuli-responsive effect when added to HNBR as filler.

Degradation of tensile properties is a common problem for swellable packers and other sealing elements. By selectively using certain classes of fly ash as effective reinforcing filler, one can develop swellable seals that can retain or even increase their modulus and stiffness upon swelling. Therefore, there is a potential to minimize or even avoid the loss of mechanical properties by water swellable reactive packing elements due to swelling. This provides an opportunity to decrease the size of these elements, which are often represented by rather long sections, and hence occupy a significant amount of space downhole. In this paper we have studied the reinforcing effects of fly ash fillers with HNBR compounds as well as stimuli responsive properties of these compounds.

EXPERIMENTAL SECTION

Rubber compounds T1, T2, T3, and T4 are elastomers based on a custom-built peroxide curing system. Formulation F2 was prepared by adding 15 wt. % of fly ash class C to compound F1, which is a commercial swellable NBR sulfur-cured compound.

T series compounds were prepared using a C.W. Brabender® internal mixer operated at 60 RPM for 20 minutes at ~160°F and then molded into sheets and cured at 340°F for 30 minutes in a hot press. Compound F2 was prepared by mixing fly ash with compound F1 in the Brabender mixer for 6 minutes at 60 RPM at ~100°F. Compound F1 is a commercial water swellable formulation that was used as received. F series compounds were cured at 300°F for 60 minutes in a hot press.

Tensile properties of rubber samples were tested using dog bone specimens, according to the ISO37 standard, at room temperature and at 350°F using an MTS® testing system.

Swelling tests in 3% KCl brine and LVT 200 oil were performed to assess the effect of fillers on the chemical resistance of elastomers. Standard one-inch rubber buttons were tested at 200°F at ambient pressure for 6 days. Thermal properties were studied using TGA Q50 and DSC Q2000 instruments.

To analyze the stimuli-responsive effect introduced by fly ash fillers, rubber buttons were aged for 10 days in 3% KCl in the oven at 200°F and then removed from the solution, dried for 30 minutes in the oven and then for 3 hours at room temperature. Samples were later tested in the MTS testing system in compression mode.

Viscoelastic properties of stimuli-responsive samples, based on water swellable compounds, were accessed using the RSA G2 dynamic mechanical analysis (DMA) instrument. This experiment was conducted in a temperature ramp mode of 2°C/minute at 1Hz frequency and 0.2% amplitude strain. A Nicolet 6700 Fourier Transform Infrared (FTIR) spectrometer was used to confirm neutralization of carboxylic groups by metal cations.

RESULTS AND DISCUSSION

The reinforcing properties of fly ash were evaluated at room temperature (Table 1) and at 350°F (Table 2) and compared with N550 carbon black. Sample T3, reinforced with fly ash, demonstrated inferior tensile properties and hardness compared to a compound filled with carbon black. However, at room temperature, fly ash was capable of delivering a reinforcing effect to the rubber compound without sacrificing its elasticity, e.g. the ultimate elongation of elastomer compound remained virtually unchanged after incorporation of 50 parts per hundred rubber (PHR) of fly ash, while its modulus and tensile strength have increased. Therefore, potentially, one can substitute carbon black with high loadings of fly ash in certain applications, thus further decreasing the cost of the rubber product. The summary of mechanical properties is provided in Table 1, E_x denotes modulus of a compound at certain strain (x), i.e. stress at strain.

Sample	N550 CB (PHR)	Fly ash Class F (PHR)	Tensile strength (psi)	E_{25} (psi)	E_{50} (psi)	E_{100} (psi)	Elongation (%)	Hardness Shore A
T1	0	0	829	71	97	121	472	48
T2	50	0	3402	165	273	744	290	68
T3	0	50	1176	94	127	166	467	54

Table 1. Room temperature tensile properties and hardness of elastomer samples

The relatively low increase in modulus and hardness of T3 compared to T1 at room temperature can be explained by a low aspect ratio of fly ash and weak interactions between blend components. The latter problem can be solved by functionalization of fly ash particles.

The general data trends from the high temperature tensile tests closely resemble the results of the tensile test conducted at room temperature. However, at high temperature, rubber specimens break at much lower elongation, which leads to a great decrease in the measured tensile strength. This behavior is typical for elastomers and limits their application to a certain maximum service temperature and is also due to the fact that high temperature tests are conducted at conditions that push the limits of the material. The modulus values at 15% (E_{15}) and 20% (E_{20}) strain are not provided herein, because half of the tested specimens were broken before they reached such strains.

The superiority of carbon black as an efficient filler became even more pronounced at high temperature. However, incorporation of fly ash doubled the tensile strength and increased the elongation at break by one-third when compared to the unfilled elastomer.

Sample	E_5 (psi)	E_{10} (psi)	E_{15} (psi)	E_{20} (psi)	Tensile strength (psi)	Elongation (%)
T1	31	46			91	23
T2	61	130	194	261	832	39
T3	42	73	98	119	173	30

Table 2. Summary of high temperature (350°F) tensile data for elastomer samples

Thermal properties of samples T1-T3 were assessed with Differential Scanning Calorimetry (DSC) and Thermogravimetric analysis (TGA). TGA allows evaluating thermal stability of samples and to verify the actual composition of a sample from the ash content. The results of the DSC and TGA analysis are summarized in Table 3. During TGA, samples were heated to 600°C at 10°C/minute rate in a nitrogen (N_2) environment, and subsequently heated at 10°C/min to 600 °C in air to burn off carbon-based fillers. This allows decomposing of organic matter first by applying heat, and then burn the remaining carbon-based material in air atmosphere.

The initial degradation of the polymer started at 200°C and progressed to complete decomposition of aliphatic chains in the temperature range of 420-500°C with a highest rate of decomposition at ~460°C. Sample T2 incurred the second rapid mass loss after 600°C due to burning of the carbon black, which comprised most of the ash content at 600°C in a N_2 environment. The DSC experiment demonstrated the glass transition temperature of the elastomer remained largely independent of the type of filler - the respective DSC heating scan was conducted with a 10°C/minute temperature ramp after precooling the sample to -50°C.

Sample	W_{600} (%)	W_{800} (%)	$T_{deg}1$, °C	$T_{deg}2$, °C	T_g, °C
T1	2.4	2.4	452.8	-	-29
T2	29.8	2.7	458.7	616.5	-29
T3	32.8	32.7	447.3	-	-28

Table 3. TGA/ DSC results for elastomer samples

Table 3 contains ash contents (W) at 600°C and 800°C, degradation temperatures (T_{deg}) which represent the temperature of the highest rate of decomposition during TGA experiment, and glass transition temperature (T_g) of the samples, as determined by DSC scan.

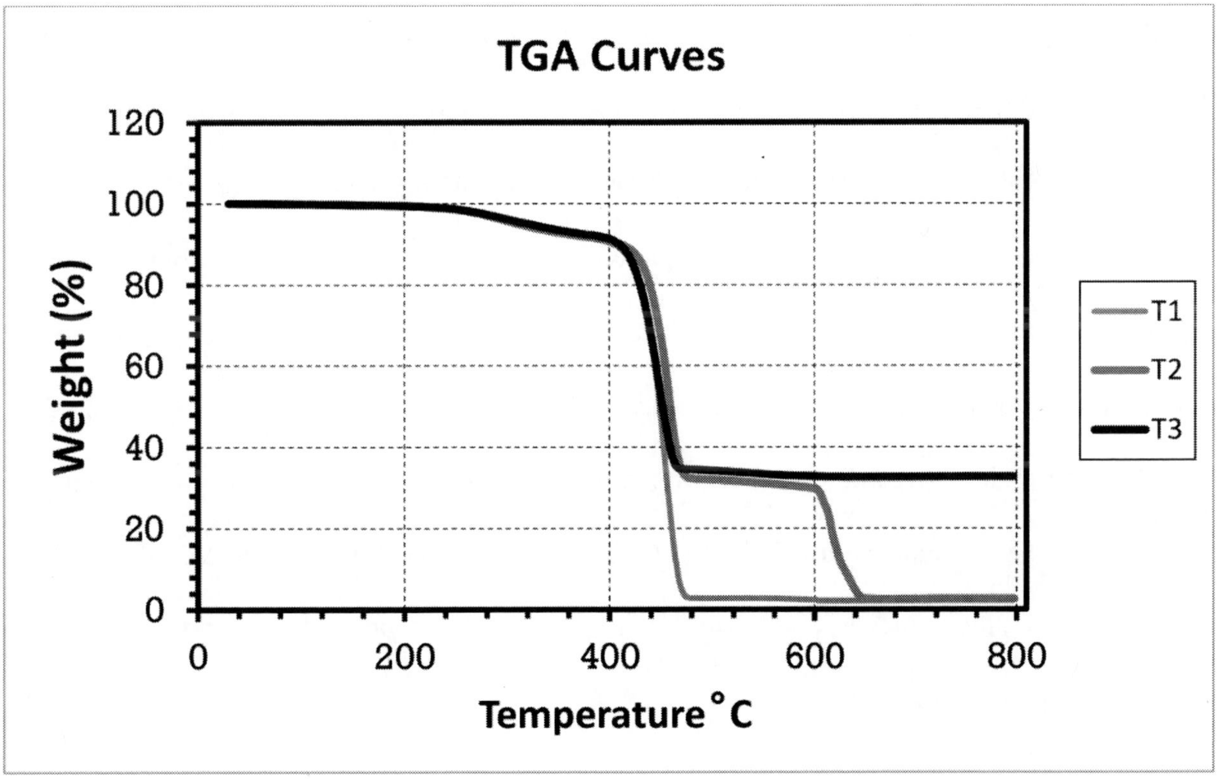

Figure 1. TGA Curves for samples T1, T2, and T3

Testing fluid / Sample	3% KCl brine	LVT 200	Specific gravity
T1	0.6	13.0	0.96
T2	0.3	11.6	1.13
T3	0.8	9.1	1.18

Table 4. Volume increase (%) of elastomer samples due to swelling for 6 days at 200°F

Musso et. al. developed a stimuli-responsive HNBR composite reinforced with Portland cement that was increasing its stiffness and modulus when swelled in water[3]. Fly ash (Class C) is a common component of Portland cement, which contains the components necessary to achieve a similar effect. Therefore, one can avoid the loss of mechanical properties by water swellable packing element seals upon activation. To test this hypothesis, we studied the effect of aging in water brines on the properties of elastomers with fly ash fillers.

Rubber samples were molded and cured into standard size rubber buttons (cylinders of ~0.25" height and ~1" diameter). These buttons were aged for 10 days in 3% KCl brine in the oven at 200°F, removed from the solution, dried for 30 minutes in the oven and for 3 hours at room temperature. The swelling of both compounds was close to ~1% by volume (Table 4). Samples were later tested in an MTS machine in compression mode. Each button was compressed to 70% strain, buttons remained intact during the experiment, and visually all appeared to have a certain level of compression set at the end of the experiment. The results of the experiments are summarized in Table 5 and Figure 2. Only one button was tested for each sample. The provided compression modulus values (C_n) represent values of stress at a percentage strain (n), e.g. C_5 is stress at 5% compression strain.

Compound T4 was reinforced by fly ash Class C, which contains larger weight fraction of CaO than fly ash Class F (~25% vs ~10-15% respectively) and is expected to produce a higher pH and closely resemble cement that was used to prepare stimuli-responsive rubber in the original paper by Musso et.al.[3] The results demonstrate that exposure to water (3% KCl) at high temperatures can improve the modulus for the compounds filled with fly ash Class C (T4), while the control sample (T1) showed the opposite behavior. After aging, the compressive modulus at 50% strain (e.g. stress at 50% strain) of control sample T1 decreased by 19%, as compared to the unaged sample, while compressive strength of sample T4, reinforced by fly ash, increased by 15%. The difference in modulus values is even more drastic at 70% compression strain (-35% and 31%, respectively). This is in agreement with the hypothesis stated earlier.

Sample	Filler	Compressive modulus (psi)				
		C_5	C_{10}	C_{25}	C_{50}	C_{70}
T1	None	19	42	127	468	1810
T1 swollen	None	16	36	109	380	1185
T4	Fly Ash class C (50 PHR)	21	46	134	434	1628
T4 swollen	Fly Ash class C (50 PHR)	29	57	153	501	2125

Table 5. Summary of compressive modulus properties after 3% KCl soaking and 200°F drying

To further investigate the stimuli-responsive property of fly ash filler, the latter was incorporated into a water swellable compound, which provided for significantly more interaction of filler with water to occur. Sample F2 comprises unmodified commercial water swellable NBR formulation F1 mixed with 15 wt.% of fly ash Cass C. Sample F2 was processed into a 0.04" slab, then a rectangular sample was cut from the slab and swollen in 3% KCl solution until 100% volume increase. The F2 sample was then dried overnight in the oven, which lead to de-swelling of the sample due to solvent evaporation. Mechanical properties of the F2 sample were assessed with DMA prior to the swelling test, in a swollen state, and after drying - the data summary is shown on Figure 3.

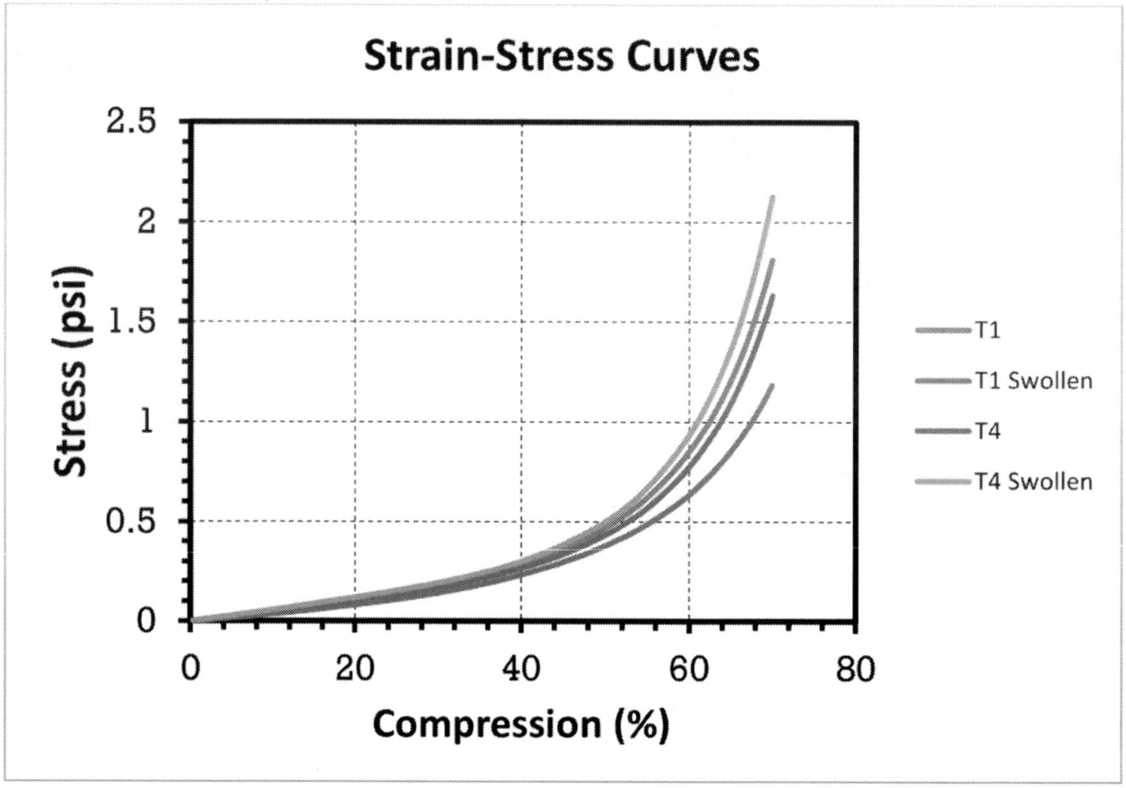

Figure 2. Stress-strain curves

The storage modulus of sample F2 after swelling and subsequent drying was about an order of magnitude higher than that of original sample F2 measured prior to swelling experiment. The storage modulus of the dried F2 sample was also less sensitive to the increasing temperature, contrary to the weakening original sample F2. The increase in the storage modulus of swollen F2 with temperature is simply due to the drying of the sample. The obtained result demonstrated that fly ash can essentially act as a stimuli-responsive material.

The significant improvement in the modulus is believed to be due to the neutralization of the swellable component of the commercial F1, and potential hydrolysis with subsequent neutralization of NBR elastomer. This is confirmed by the shift in T_g to a lower temperature range after drying as illustrated in Figure 3.

Carboxylic groups were detected in the original sample F2 prior to swelling, hence, they cannot be attributed solely to hydrolyzed HNBR, but rather to the super-absorbent polymer present in water swellable rubber compound (see Figure 4). FTIR confirms neutralization of carboxylic groups because a significant increase of the peak near ~1540 cm^{-1} and a lowering of peak intensity at ~1720 cm^{-1} have been observed. This neutralization is believed to promote the formation of ionic clusters in the compound with fly ash considered to be the source of cations. The neutralization of carboxylic groups could as well be attributed to K^+ cations present in the KCl brine solution, however, in that case, the respective peak in FTIR spectra would appear at higher wave numbers, approaching ~1600 cm^{-1}.

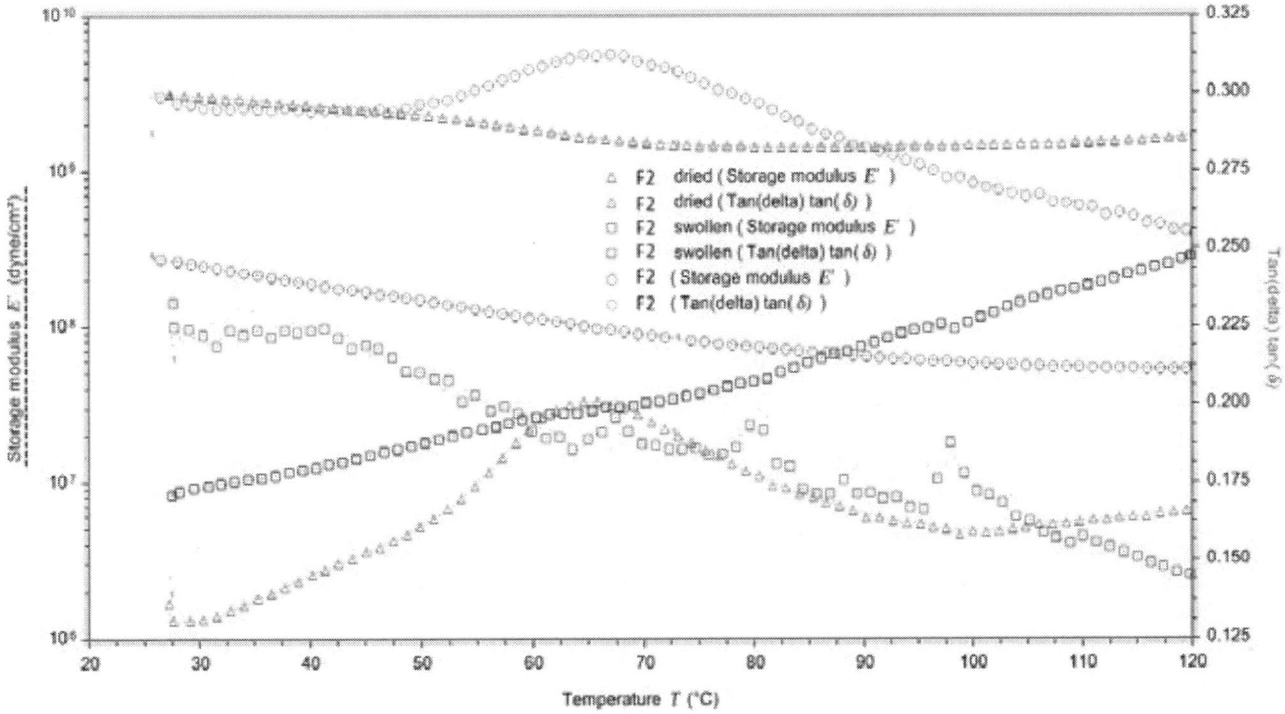

Figure 3. DMA data on F2 sample

It is important to mention that sample F2 exhibited a certain degree of brittleness after drying, which is undesired in rubbers, but this can be managed by incorporation of common ionomer plasticizers.

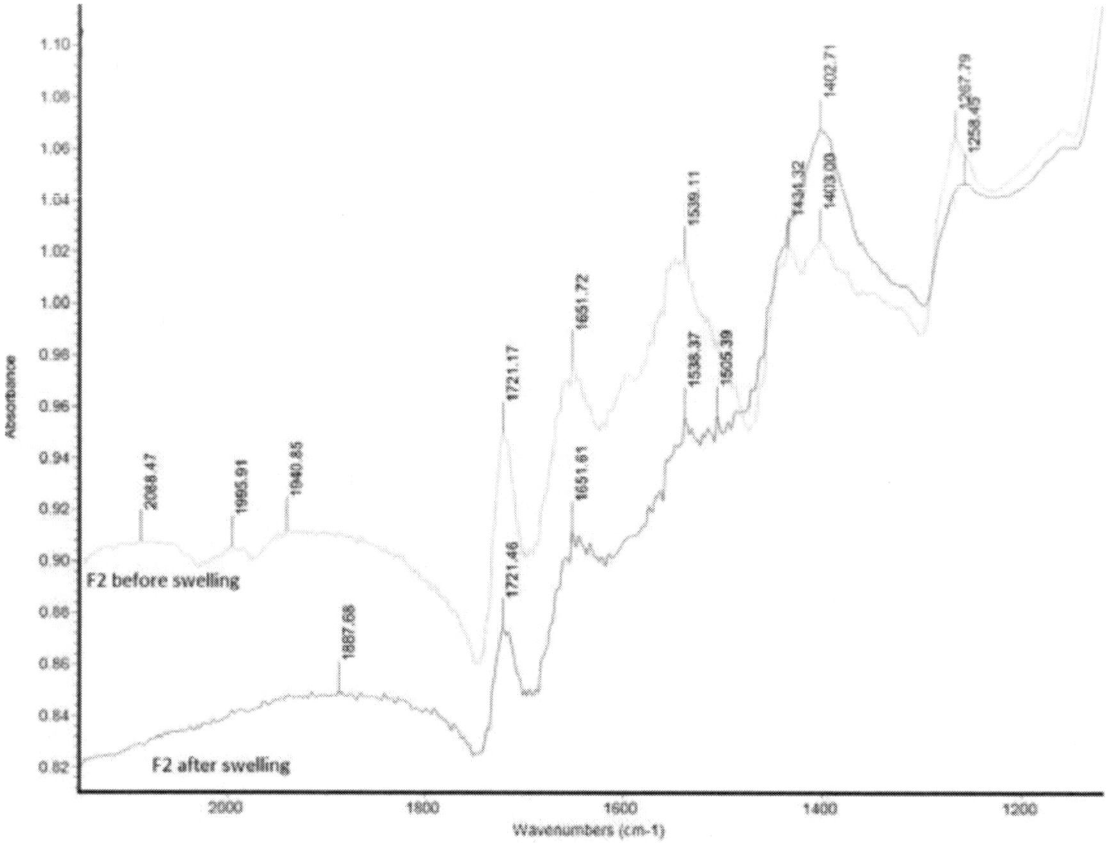

Figure 4. FTIR spectra of F2 sample taken before and after swelling

Conclusions

Fly ash demonstrated a great potential for consideration as cost effective, low reinforcing filler for downhole oil and gas applications. Although the reinforcing properties of fly ash are inferior to carbon black, incorporation of fly ash can improve the mechanical properties of HNBR elastomer compared to the unfilled raw crosslinked elastomer at both room temperature and at 325°F.

Furthermore, composites of HNBR with fly ash Class C demonstrated a stimuli-responsive effect by significantly increasing their modulus upon swelling in water brine. This effect was even more pronounced in composites of fly ash with swellable water compounds, which provided an easy access of filler to water upon swelling. A ten-fold increase in modulus was achieved in such fly ash reinforced sample, after swelling in 3% KCl water brine and drying in the oven overnight. The FTIR confirmed neutralization of carboxylic groups present in the investigated system, which contributed to overall improvement of mechanical properties of the compound upon swelling.

444310 Formal Modeling of Accidents and Automatic Generation of Tutoring Scenarios for Intelligent Safety Training of Process Operators

Abstract:

With the emphasis of accident-free operations and prevention of catastrophic events, safety-training demands for the chemical plants have been on the increase globally. However, the research results on modeling of accidents and involving human operators and the advanced theory and technology for (safety) operator training have been shared limitedly among domain experts and commercial developers. Even with the recent trend of simultaneously delivering the operator training system with the construction of a new process plant, in commercial operator training systems (OTS), safety training used to implement quite limited and obvious scenarios (and the burden was almost up to the trainer), partially due to the difficulty in modeling propagations of probable accidents. By the regulatory requirements or proactive investment by companies, effective scenarios for training of safety and abnormal situation management are necessary to be improved in the aspects of the size of volume as well as the training depth. Thus, considering the emergence of enhanced virtual reality technologies and inexpensive gadgets like HMD, supporting the immersive training environment, the future efficiency of the training heavily relies on the quantity and quality of safety scenarios, which would be essential in guiding the entire training process.

In this research we first overview a government-sponsored research project developing an enhanced safety training system for chemical process operators, using haptic devices in addition to enhanced audio and visual equipment. The initial version of the system runs with a fixed set of training scenarios developed for predetermined accident scenarios. However, with a limited set of training scenarios, the trainees simply memorize and rote learn the steps, required to be mastered to proactively solve the problems with an advent of possible accidents. In addition to that, without formal modeling, the system cannot give timely feedbacks and the performance of the trainee cannot be measured in real-time. Then to enhance the contents and usefulness of the training, the knowledge of process operations is collected and systematically stored as a form of knowledge base, in addition to the accident scenarios obtained by executing Fault Tree Analysis and HAZOP study.

By using a formal modeling language like UML and Petri nets, the accident scenarios and operators' problem-solving tasks are modeled hierarchically and classified into distinguishable subtasks. Then a new training scenario is generated on-line through automatic synthesis based on planning using the knowledge base. Safety training scenarios are generated intelligently for the tutoring of a single operator's job function as well as effective collaboration among a group of operators for real-scale malfunction diagnosis and intervention at the initial propagation stage of an accident. As case study, safety trainings for city gas governing stations are executed based on systematic scenario generation and analyzed for improved effectiveness and field applicability of the system. Reduction of time and cost in the training scenario production and real-time customized training based on new operator's level of knowledge and skill are also discussed in the framework of intelligent tutoring systems.

444376 Ozone Source Apportionments of Local Anthropogenic Emissions

Abstract:

Ozone is one of six principal air pollutants designed by US EPA. NOX and VOC emissions from anthropogenic sources are the most important factors to control ozone pollutant.

In this report, photochemical dispersion models are used to estimate the effect of ozone from local anthropogenic emission sources. EKMA-type diagrams are generated for NOX- and VOC- effects. Furthermore, APCA (Anthropogenic Precursor Culpability Assessment) technology is used to estimate ozone source apportionments.

EKMA-type diagrams show that 10-30% emission reductions could reduce ozone concentrations to meet standard. Further APCA results show the average ozone contributions from regions and emission categories. APCA results explain different effects for different monitoring stations.

444397 Flare Minimization and Emission Characterization during Start-Ups of Chilling Train in Ethylene Plants

Abstract:

Flaring is a crucial safety operation to protect chemical plant personnel and equipment during plant start-up. However, it generates huge amounts of off-spec products, and causes negative environmental and social impacts, as well as results in tremendous raw material and energy losses which could be saved to generate much more needed products. Thus, cost-effective start-up flare minimization strategies which is a double-win practice, simultaneously benefiting both the industrial and environmental sustainability, are becoming more and more important and attractive to the industries.

The chilling train is a very important section for the ethylene plants and accounts for more than 50% of the total start-up time and flare emissions of the whole ethylene plants. Within this section, the charge gas from the upper stream process is refrigerated to a very low temperature (lower than -160 °C) to separate hydrogen and methane. Meanwhile, the left charge gas is liquidized for cryogenic separation. The chilling train system consumes over 40% of the refrigerating capacity of an ethylene plant. During the chilling train start-up, different operational conditions with different compositions and flowrate of precooling process will affect both product separation efficiencies, start-up stability time and flare emission.

Traditional start-up scenarios are developed based on experience and improved by trial and error, which is inefficient and dangerous. In this study, a general methodology on flare minimization for chilling train start-up operations via rigorous dynamic simulation is presented. The methodology starts with setup and validation of chilling train steady-state and dynamic simulation models. The initial state of start-up in the real plant is achieved through systematically transforming from the validated dynamic model, and several operation conditions with different compositions and flowrate of precooling stream are examined to identify an optimal initial state of start-up based on the obtained initial state. Thereafter different flare minimization strategies for the chilling train start-up are conducted on the optimal initial state model. Also, emission source characterization under different flare minimization strategies for chilling train start-up is investigated. Deep insights of the start-up strategies, emission source distribution and dynamic emission profiles are provided. This study provides detailed technical support for both the industry and environmental agencies on evaluating and developing cost-effective flare minimization strategies in the future.

Experimental Study on Najas Minor of Producing Bio-Oil

Abstract: Due to the diminishing of petroleum reserves and the environmental degradation caused by consumption of fossil energy , many attentions have been paid to the application of biomass energy which is acceptable as an abundant renewable energy resource. Najas minor grows in dense clusters with highly branched stems of which a small fragment could propagate into a thrive one.In this study, we researched the properties of Najas minor to make biofuels .Firstly,the najas minor after heavy metal ions degradation were fed into pyrolyser to prepare biofuels. Significant differences of the product was observed compared with only Najas minor pyrolysed.

KEY WORDS: Najas Minor; catalytic pyrolysis; bio-oil

1.Introduction

Energy demands and environmental problems caused by fossil energy have been a matter of growing concern.Bio-oil has been considered to be a promising way for producing renewable bio-fuels or chemicals[1-3].In recent years,many countries pay attention to the importance of biomass that is an abundant renewable energy resource.Biomass can be defined as any organic material mainly consisting of carbon, hydrogen,oxygen, nitrogen, and some other components in small proportions[4].Although the biomass has been studied for nearly a century,most studies mainly concentrated in Lemna minor, Lemna gibba, Spirodela polyrhiza and Landoltia punctata.We still know little about the biological characteristics and potential application value of other varieties of biomass[5].Najas minor,a submersed herb,grows in dense clusters with highly branched stems of which a small fragment could propagate into a thrive one.Najas Minor is native to Europe, Asia and North Africa from the Netherlands to Morocco east to Japan and the Philippines,including China,Siberia, Central Asia, Iran, Turkey, Ukraine, Germany, France Italy and a host of other Countries.

Catalysts used for catalytic cracking are basically divided into the following categories: molecular sieve, alkali metal salt, metal oxide, etc.Different catalysts have different catalytic effects.The use of metal nanoparticles has shown to be very important in recent industrial applications[6].The application of precious metal nanometer materials in catalysis,which has caused the extensive concern of researchers in recent 30 years,has become a hot research topic in the field of catalysis.Gold nanoparticles have received a great deal of attentions due to their special optical and electrical performances, catalytic properties and remarkable quantum size effect, surface effect etc[7]. The synthesis of metal nanoparticles can be classified broadly into wet and dry methods[8]. The cost of production is relatively high both materially and environmentally. There is an unequivocal need to develop more cost-effective and environmental benign methods to alternate these existing methods[9].Researchers in this field turn to use organisms to produce inorganic nanostructures extracellularly[10].

Although there has been a large volume of work on the synthesis of gold nanoparticles[11-13], and obtainning bi-oil by abundant biomass, there are relatively fewer attempts to combine these two methods.In this work, gold nanoparticles were obtained by aqueous chloroauric acid solution with the of Najas Minor at room temperature. The direct pyrolysis and catalytic pyrolysis experiments of Najas Minor were carried out to observe the effects of catalysts on pyrolysis process of Najas Minor.

2. Materials and Methods

2.1. Experimental samples

The Najas Minor collected from Tianjin city and the solutions of Au(III) were used in this study.The stock solutions of Au(III) were prepared by dissolving analytical grade chloroauric acid

($AuCl_3 \cdot HCl \cdot 3H_2O$, Kewei Reagent, China) in deionized water.

2.2.Experimental apparatus and procedure

Pyrolyzer with GCMS,shown in Fig. 1,is mainly used for analyzing the composition of polymer.Pyrolyzer with GCMS includes three parts :quartz pyrolysis tube,pyrolyzer and GCMS.

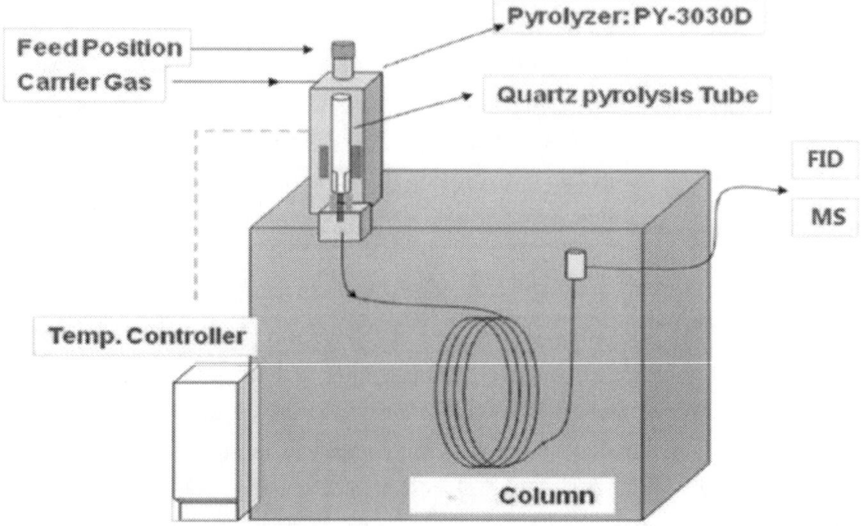

Fig. 1 Diagram of Pyrolyzer with GCMS

The quartz pyrolysis tube (shown in Fig.2) length is 12cm.The sample end diameter is 5 mm,and the pyrolysis products end diameter is 3mm.This special structure is beneficial to the fixation of the reactant in the tube.The top and bottom of the reactants are both supported by quartz wool to prevent reactants blowed by the carrier gas ,or fine particles into the chromatographic column.All pyrolysis experiments were carried out on the EGA/PY-3030D pyrolyzer(Frontier Lab,Japan) .The samples were filled in the special quartz pyrolysis tube to reduce the amount of valuable catalyst and reduce the cost of research work.

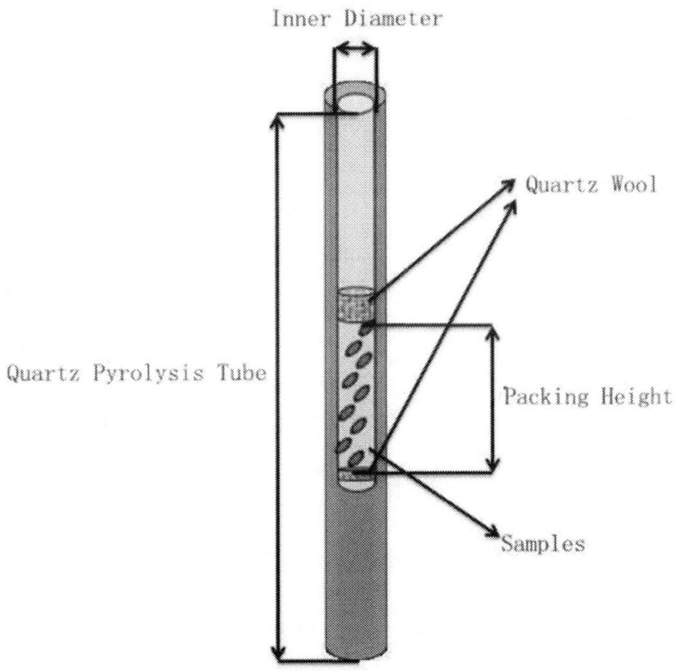

Fig. 2 Diagram of Quartz Pyrolysis Tube

Approximately 0.5g Najas Minors were used in each experiment.The samples were impregnated in 50ml deionized water and 50ml solution of chloroauric acid(0.0793mmol/L) for a night,followed by drying at 80℃ until the weight didn't change.Each experiment used 2.5mg Najas Minors as reactants.Each sample was heated, under the control of heating controller, from ambient temperature to 600 ℃ repeatly and kept at 600 ℃ for 2 h.

3. Results and discussion

3.1.MS Analysis of the Products during the Pyrolysis

Under the same pyrolysis conditions, bio-oil obtained by direct pyrolysis and catalytic pyrolysis were analyzed by GCMS,and the main chemical compounds were shown in Table1 and Table2.

Table 1 Indentification and quantification of main chemical compounds by direct pyrolysis

NO.	peak area%	Formula	Compound
1	12.67	$C_3H_6O_2$	2-Propanone, 1-hydroxy-
2	3.67	C_7H_8	Toluene
3	3.28	C_7H_8O	p-Cresol
4	3.81	$C_5H_{10}O$	Pentanal
5	3.53	C_8H_8O	Benzofuran, 2,3-dihydro-
6	2.17	C_8H_8N	Indole
7	5.93	$C_{20}H_{40}O$	3,7,11,15-Tetramethyl-2-hexadecen-1-ol
8	2.52	$C_{22}H_{42}O_2$	Phytol, acetate
9	5.25	$C_{16}H_{32}O_2$	n-Hexadecanoic acid
10	8.99	$C_{18}H_{30}O_2$	9,12,15-Octadecatrienoic acid, (Z,Z,Z)-

Table 2Indentification and quantification of main chemical compounds by catalytic pyrolysis

NO.	peak area%	Formula	Compound
1	6.01	C_7H_8O	p-Cresol
2	1.75	$C_7H_7NO_2$	3-Hydroxypyridine monoacetate
3	4.58	C_8H_7N	Indole
4	3.05	$C_{10}H9N$	1-Phenyl-1-cyclopropanecarbonitrile
5	3.90	$C_{12}H_{22}O_{11}$	.alpha.-D-Glucopyranose, 4-O-.beta.-D-galactopyranosyl-
6	8.04	$C_{20}H_{40}0$	3,7,11,15-Tetramethyl-2-hexadecen-1-ol
7	2.82	$C_{20}H_{40}$	2-Hexadecene, 3,7,11,15-tetramethyl-, [R-[R*,R*-(E)]]-
8	3.99	$C_{15}H_{28}$	1-Pentadecyne
9	2.12	$C_{15}H_{28}$	3,7,11-Trimethyl-2,4-dodecadiene
10	8.75	$C_{16}H_{32}$ 0_2	n-Hexadecanoic acid

3.2 discussing

According to the chemical category and common characteristics, the products were divided into hydrocarbons, alcohols, acids, ketones, nitrogen compounds and other compounds.Nitrogen compounds detected by GC/MS analysis ,mainly come from the pyrolysis of amino acids in Najas Minor.Table 3 showed the distribution of pyrolysis products.As shown in the table,the hydrocarbons ,alcohols and nitrogen compounds in biological oil of catalytic pyrolysis were significantly increased compared with direct pyrolysis.The content of hydrocarbons increased from 12.4% to 34.6%.The content of alcohols increased from 9.1% to 16.6%.The gold nanoparticles loaded in surface of Najas Minor, was conducive to the generation of hydrocarbon and alcohol in pyrolysis.By analyzing the GC/MS results, the main hydrocarbons in bio-oil by catalytic pyrolysis were long chain alkanes,and olefins.The main components in gasoline and diesel is hydrocarbons. As a result, whether the biomass can be effective as future power fuel oil, depends largely on the hydrocarbon content of bio-oil after further treatment.

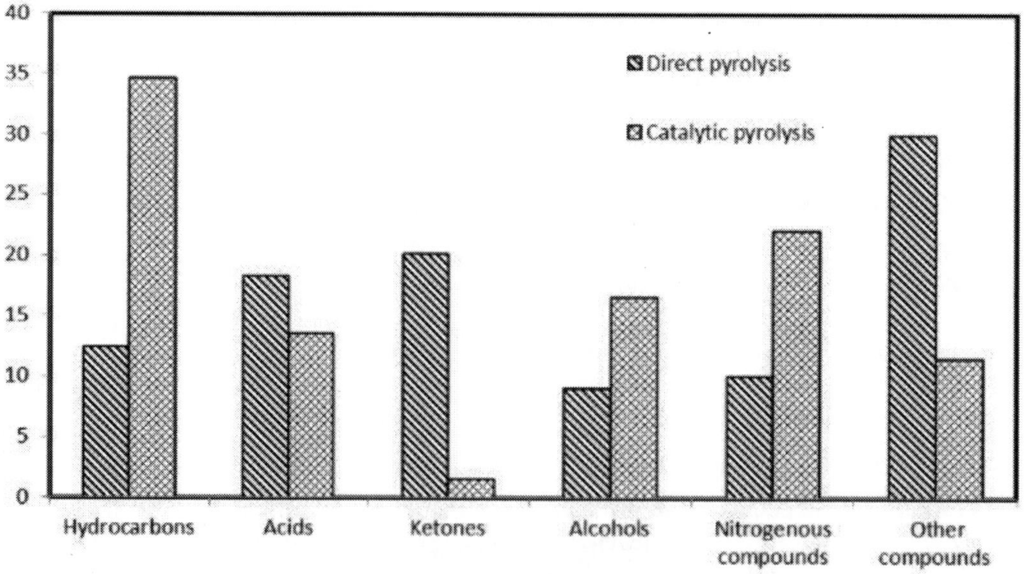

Fig.3 Distribution of pyrolysis products

4.Conclusion

Biomass was explored as a potential biomass energy resource. The experimental study on the direct and catalytic pyrolysis of Najas Minor had been presented, using Pyrolyzer with GCMS,to observe the catalytic effects of gold nanoparticles.In this study,remarkable catalytic effects for pyrolysis were observed.The pyrolysis of Najas Minor with gold nanoparticles,provided more hydrocarbons, alcohols and nitrogen compounds.

参考文献：

[1]Gong F, Yang Z, Hong C, et al. Selective conversion of bio-oil to light olefins: Controlling catalytic cracking for maximum olefins[J]. Bioresource Technology, 2011, 102(19):9247-54.

[2]Zhang Z, Bi P, Jiang P, et al. Production of gasoline fraction from bio-oil under atmospheric conditions by an integrated catalytic transformation process[J]. Energy, 2015, 90:1922-1930.

[3]Hui S, Yuqi D, Jinzhao D, et al. Transesterification of sunflower oil to biodiesel on ZrO2 supported La2O3 catalyst.[J]. Bioresource Technology, 2010, 101(3):953–958.

[4]Yaman S. Pyrolysis of biomass to produce fuels and chemical feedstocks[J]. Cheminform, 2004, 35(31):651-671.

[5]Hejný S. Ellias Landolt The family of Lemnaceae—A monographic study—Vol.1 biosystematic investigations in the family of Duckweeds (Lemnaceae) (vol. 2)[J]. Folia Geobotanica Et Phytotaxonomica, 1993, 28(1):50-50.

[6]Gardea-Torresdey J L, Tiemann K J, Gamez G, et al. Gold Nanoparticles Obtained by Bio-precipitation from Gold(III) Solutions[J]. Journal of Nanoparticle Research, 1999, 1(3):397-404.

[7]Marie-Christine D, Didier A. Gold nanoparticles: assembly, supramolecular chemistry, quantum-size-related properties, and applications toward biology, catalysis, and nanotechnology.[J]. Chemical Reviews, 2004, 35(16):293-346.

[8]Liu B ,, Xie J ,, Lee J Y, et al. Optimization of high-yield biological synthesis of single-crystalline gold nanoplates.[J]. Journal of Physical Chemistry B, 2005, 109(45):15256-63.

[9]Raveendran P, Fu J, Wallen S L. Completely "Green" Synthesis and Stabilization of Metal Nanoparticles[J]. Journal of the American Chemical Society Jacs, 2003, 125(46):págs. 13940-13941.

[10]Absar Ahmad , Satyajyoti Senapati , MIK, et al. Extracellular Biosynthesis of Monodisperse Gold Nanoparticles by a Novel Extremophilic Actinomycete, Thermomonospora sp.[J]. Langmuir, 2003, 19(8):3550-3553.

[11]Selvakannan P, Mandal S, Phadtare S, et al. Water-dispersible tryptophan-protected gold nanoparticles prepared by the spontaneous reduction of aqueous chloroaurate ions by the amino acid[J]. Journal of Colloid & Interface Science, 2004, 269(1):97-102.

[12]Gittins D I, Caruso F. Biological and Physical Applications of Water-Based Metal Nanoparticles Synthesised in Organic Solution[J]. Chemphyschem, 2002, 3(1):110 - 113.

[13]And A G, Murphy C J. Seed-Mediated Synthesis of Gold Nanorods: Role of the Size and Nature of the Seed[J]. Chemistry of Materials, 2004, 16(19):3633-3640.

444353 A Combined Rheology and Stability Study on Emulsified Acid for High Temperature Applications

Abstract:

HCl emulsion has been a popular formulation to stimulate carbonate reservoirs due to the various advantages that emulsified HCl has over ordinary HCl; including but not limited to deeper penetration and less corrosion damages. Emulsified HCl has even more potential advantages if utilized properly. The reaction kinetics, stability and rheology of emulsified HCl plays a significant role on achieving successful acidizing treatment.

The objective of this paper is to experimentally study the factors affecting the stability of emulsified HCl. The emulsion was prepared using 15wt% HCl and diesel with cationic emulsifier. The acid solution-to-diesel ratio was 70:30. The effect of stirring intensity, concentration of the emulsifier, addition of corrosion inhibitor, temperature, addition rate of acid phase to oil phase and mixing time of the emulsion have been studied. Stirring intensity was varied from 600 to 1500 rpm. Emulsifier dosage ranged between 5 to 20 gpt. Temperature used for heating, to mimic reservoir condition, was as high as 120°C.

The results showed that there is an optimum condition for emulsifier dosage and for mixing time of the emulsion. The optimum emulsifier dosage was found to be 10 gpt with 5 minutes of mixing time. As for the addition rate of acid phase to oil phase, 1.4 ml/min of drop-wise addition was found to give a very stable emulsion. The effect of stirring intensity on the emulsion stability was proportional, the higher the rpm speed the better the stability. Corrosion inhibitor, on the other hand, played a role as a destabilizer. With regards to heating temperature, it was found out that the higher the temperature, the faster the emulsion breaks.

The novelty of this study is the comprehensive approach in combining stablity and rheology of emulsified acid for high temperature applications. The particle size analysis and solution morphology help in better understanding HCl emulsion and how it works.

COMPARATIVE ANALYSIS OF MOLECULAR STRUCTURE IDENTIFIABILTY BASED ON SIGNATURES AND DESCRIPTORS

Zelimir F. Kurtanjek
University of Zagreb, Croatia

Abstract

This work is focused on molecule computer representation aimed for molecule predictive model (QSAR) and inverse modelling. Two different approaches are compared, i.e. the molecule structure coding based on graph theory (Faulon et al.[1] extended valence) and the chemical based descriptors[2]. Applied are 1-3rd level molecule graph fragments and 630 2D molecular descriptors. The molecule test sets are alkanes, alkenes, acetones, aromatics, organic acids and halogenated hydrocarbons in the range of C_1-C_{12}, selection of biomolecules (including aminoacids) and a set of binary ionic liquids (cations: imidazole, pyridinium, quinolinium, ammonium, phosphonium). The comparative comparison of molecule description based prediction potential are statistically evaluated by octanol-water (PWA) partition coefficients (for hydrocarbons), viscosity of ionic liquids and biological impact on acetycholinesterase inhibition. The evaluation is based on two distinct levels. On the first level the comparative analysis based on internal and external correlation between "scores" in multidimensional spaces of the separate first, second and third order valance spaces, and the molecular signature space. Firstly, the potential of information content applicable for inverse modeling is tested on ability for molecule classification by use of linear discrimination (LDA) based on the singular value decomposition (SVD) of corresponding molecular fragments and descriptors. For chemical-biological property inverse modelling proposed is application of constrained genetic algorithm (GA). Tested are linear structure multivariate (chemometric) models such as principal component regression and partial least squares and nonlinear decision trees (random forest). Applied are singular value decompositions for reduction of the molecule signature and the descriptor spaces up to the significance level of p=0.05. The selected level of significance is presumed to be in coherence with experimental data of molecule physical and chemical properties. The main focus is on extraction of the main molecular fragments (partial graphs extracted from molecular signatures) and specific molecule descriptors. Importance of application of open source software modules provided by Bioclipse[3] and R software packeges[4] for computational-statistical numerical algorithms is emphasized.

Key words: Molecule signatures, descriptors, inverse modelling, decision tree

Introduction

Advancement of high-throughput experiments have lead to „big data" databases applied in pharmaceuticals, medicine, new material science, catalysis, and new industrial processes. The role of chemoinformatics as a science for computer encoding of molecule properties and algorithms for data mining and inference are the essential steps in the development. Computer encoding of molecular properties has been a part of chemical engineering from the beginning of process design. It is mostly based on molecule group contributions, but has been extended by application of molecular descriptors (1-2-3 D), molecule signatures, and application of molecular quantum mechanics. Modeling is faced with the problem of „high dimension curse". For example, about 5 000 various molecular descriptors can be determined for each molecule, although most of them are highly correlated. It results in an ill-conditioned problem of modelling with large uncertainties in parameter and poor model prediction accuracy outside modelling training sets. Great improvement has been achieved by application by use of model ensembles, like random forest of many individual decision trees. These models are highly flexible, can be trained on discontinuous and nonlinear functional relations, can encompass various data formats like numerical and categorical variables, and due to large number of randomly selected individual predictors in principle do not have tendency to overfitting. However, when the object of modelling is extraction of important molecular features, due to autocorrelation between descriptors random forest, produced lists of importance are large sets of poorly distinguishable variables. The focus of this work is to apply feature regularization for optimization and reduction of list of variable importance. Compared are two methods for structure encoding, molecular descriptors and molecular signatures of first and second order.

Modeling and Factor Regularization

In this work are selected two distinct data sets of organic molecules from open source data bases (J. Sangster, 1989) and ionic liquids ILs (The UFT/ Merck and NIST, 2015), and lists of the selected molecules are given in Table 1. As target properties for organic molecules are selected the corresponding molecular classes and partition octanol-water (POW) coefficients, and dynamic viscosity in the range of 30-80 ^{0}C of ILs is selected The molecules are coded into SMILES and used as input lists into software for evaluation of molecular signatures. Most popular open source software tools are DRAGON (R. Todeschini, et al., 2009) and PADEL (Chun Wei Yap, 2010). Here are used 1-2 D descriptors generated by PADEL. For the same list of molecules are determined first and second order molecular signatures or extended valances (J.P. Faulon, 2003). The signatures are two dimensional graph based structural molecular fragments. The vertices are atoms and the edges are corresponding bonds (lengths or heights), and are here only used of the first and second level. A molecule is characterized by a complete set of signatures as input data for modelling. Since the models are based on large number of variables (molecular features) possibly lager than number of samples (molecules), usually denoted as p>>n models. They are intrinsically „ill conditioned" leading to good predictions on train data but very poor predictions on test data. Also, produced list of variable importance are very dense "clusters are not distinguished" with large uncertainties (broad confidence intervals). This makes the problem of inverse modelling aimed to predict a set of candidate molecules for given set of target properties, such as in a case of high dimension combinatorial task for selection of ILs cations and anions, very unreliable. In order to reduce model dimensionality and improve robustness and parsimonious of important molecular features here is applied modelling with feature regularization. The model is based on ensemble of random trees (CART models, L. Breiman et al. 1998) with included feature regularization.

Data sets of organic molecules		**Range of number of atoms**			
Molecular classes	C	H	O	Cl	F – I - Br
Alkanes	5-12	12-30			
Alkenes	4-9	6-18			
Acetones	6-24	6-26			
Halogenated hydrocarbons	3-10	6-22	1		
Organic acids	1-16	2-32	2		
Esters	4-11	6-22	2		
Aromatics	1-12	1-21		3	2 – 2 - 1
Data sets of ionic liquids (ILs)					
Cations		**Anions**			
1,3-Dimethylimidazolium (+)		PF_6 (-) Cl (-) CH_3COO (-)			
1-Methylpyridinium (+)		BF_4 (-) $MeSO_4$ (-) $EtSO_4$ (-)			
1,1-Dimethylpyrrolidinium (+)		Tf_2N (-) CF_3SO_3 (-)			

Table 1. The molecular groups of the organic and ionic liquid molecules selected for random forest models and extraction of the most important molecular features. The properties for the organics are obtained from J. Sangster (1989) and for the ionic liquids from Merck and NIST data bases (2015). Molecular descriptors and signatures of ILs ions are recorded separately and adjoined as a single record.

The modelling is formulated as an optimization (functional and multivariable minimization) problem with two parts of the objective function (T. Hastie et al., 2009).

$$\min F_{object.}(X,T,\omega) = SSE(y,\hat{y}) + \phi(T,\omega) \tag{1}$$

The first part is the sum of square errors (SSE) based on difference between observed y and prediction $\hat{y}$ based on input data by an ensemble T of interrelated CART models (L. Breiman et al. 1998), The second part is a selected metrics ϕ for the model complexity depending the sum of leaf probabilities for population of individual CART models and their total number N_T .

$$\phi(T,\omega) = \gamma \times N_T + \frac{1}{2} \times \lambda \times \sum_i^{N_T} \omega_i^2 \tag{2}$$

The ensemble output prediction is a sum of predictions of individual T_i tree CART models:

$$\hat{y} = \sum_i^{N_T} T_i(X) \tag{3}$$

A very efficient open source code Extreme Gradient Boosting XGB (T. Chen, 2016) is available in R language (R Developm-ent Core Team, v. 3.2.3, 2015). The optimization algorithm is consecutive, starting with first CART model which is appended with a next tree boosted by errors in predictions of previous tree. At each step optimization is solved by linearization of the objective function to quadratic terms. Number and structure of trees as well as the parameters for regularization need to be found by trial and error and minimal number of trees by folding and cross validation error on the training set. The critical step is validation on untrained set to have acceptable model error and avoid model overfitting with minimal model complexity.

Results and Discussion

Tested are models for prediction of the seven molecular classes (Table 1.) based on molecular descriptors and molecular signatures (first and second level). Results for molecular descriptor feature importance are shown on Fig. 1. The effect of regularization is very pronounced. Out of hundreds of descriptors for the molecular classification needed are only few, the most important being BCUTw.1h. On contrary, when an ensemble of independent CART trees is applied, i.e. random forest model, level of accuracy of prediction is the same, but the list of variable importance is densely populated with numerous descriptors.

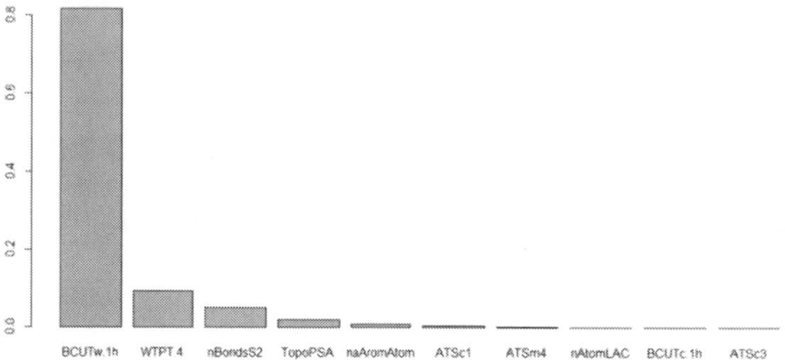

Fig 1. Variable importance molecular descriptor factors for the inverse modeling of the selected molecular classes. The first five descriptors are: BCUTw.1h (based on a weighted version of the Burden matrix which takes into account both the connectivity as well as atomic properties of a molecule), WTPT.4 (Randić's weighted path), nBondS2 (bond count), and naAromAtom (number of aromatic atoms).

The same classification problem was tested by the first and second order molecular signatures. The important signatures are depicted on Fig. 2-3. Accuracy of prediction on trained sets of molecules compared to descriptors is about the same, though the importance factors are not as distinctive as in case of the descriptors.

significantly better when molecular descriptors are applied. The

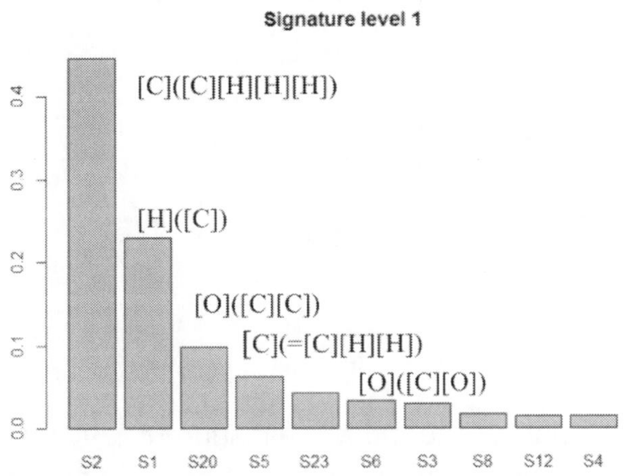

Fig 2. Relative importance of the individual first level signatures for inverse prediction of the selected molecular classes. The first five most important signatures are coded according to SMILES (Simplified Molecular Input Line Entry System).

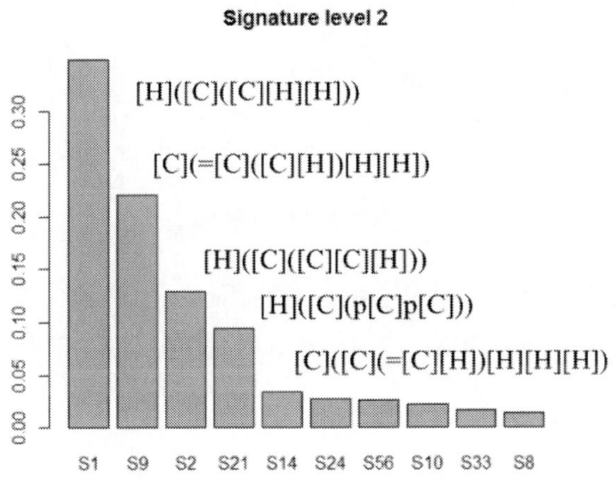

Fig 3. Relative importance of the individual second level signatures for inverse prediction for the selected molecular classes. The first five most important signatures are coded according to SMILES (Simplified Molecular Input Line Entry System).

However, model accuracy on untrained set of molecules, depicted in Fig. 4, is significantly lower for the model with molecular descriptors compared to signatures. The average error with descriptors is about 5% while for signatures about 11%. The "box and whiskers" plots are obtained by error evaluation of ten randomly generated untrained sets.

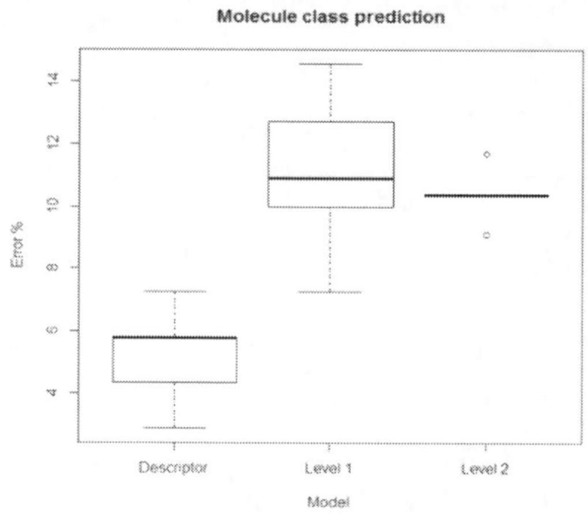

Fig 4. "Box and whiskers" plot for the model validation (relative errors) for inverse prediction of the molecular classes based on the set of untrained molecular descriptors and the first and second level of molecular signatures.

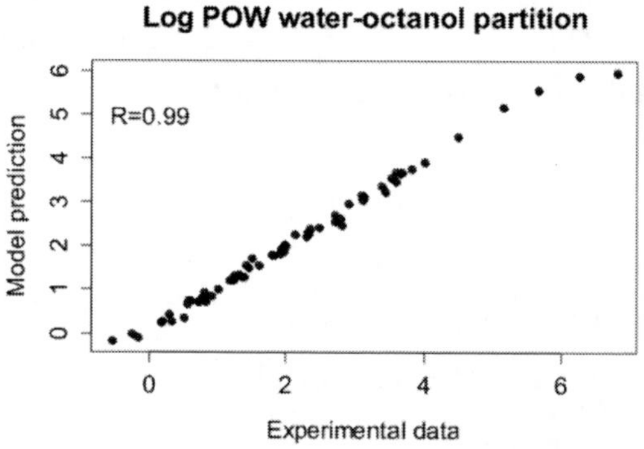

Fig 5. Predictions of the partition octanol-water coefficients POW, for the specified molecular classes of organic molecules, are based on the random forest model with the molecular descriptors.

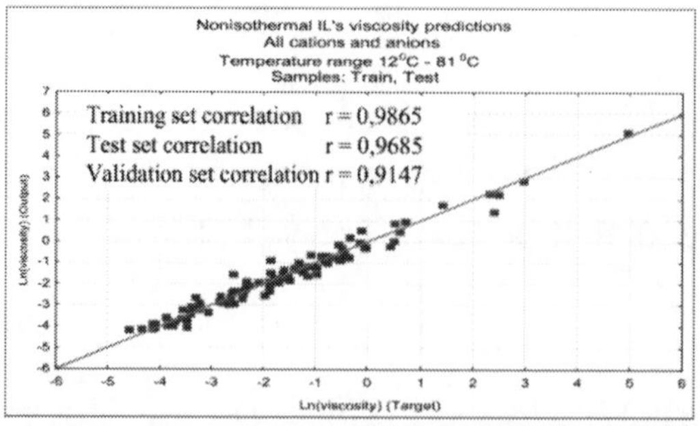

Fig 6. Predictions of dynamic viscosities for the specified groups of ionic liquids (ILs) based on the random forest model with the molecular descriptors. The viscosities cover the temperature range from 12 – 81 ^{0}C and five orders of magnitude.

In Fig. 4 can be observed that dispersion of errors (quantiles) for the second level molecular signatures is negligible, though two outliers are present.

For prediction of molecular properties applied are random forest models based on molecular descriptors. For all of the organic molecules, included in the seven classes, evaluated are octanol-water partition coefficients POW, shown in Fig. 5. The range of POW values is very high, on log scale fro 0 to 6, but the obtained accuracy of the model predictions is high (Pearson correlation R=0.99). The same approach is applied for prediction of ionic liquid viscosity data in temperature range from 12 to 81 ^{0}C. Since temperature effects on viscosity are intrinsically nonlinear, they represent a modelling challenge. However, the decision trees are very efficient for accounting nonlinearities, though nonlinearity is reflected in non-orthogonal decision planes, rendering trees with significantly increased number of branches. For the range of 6 order of magnitude the accuracy of prediction on trained set is R = 0.98 and on untrained set R = 0.91.

Conclusions

Compared are ensemble decision trees models for molecular feature class and property prediction based on extensive list of molecular descriptors and first and second level of extended valances or signatures.

The aim was to test effects of regularization on ensemble of CART decision tree models. Proven is effectiveness of the extreme gradient boosting algorithm in cases with extensive list of (molecular) features.

Obtained is increased accuracy on untrained sets of data, and more importantly, obtained are parsimonious lists for ranking of variable importance lists. This enables further significant reduction in development of robust and simple decision tree models.

The reduction in number of molecular features applied for decision tree models may greatly simplify the challenging problem of inverse modeling based on molecular descriptors and/or signatures.

References

1. J. Sangster, J. Phys. Chem. Ref. Data, **18**(3)(1989) 1111-1227
2. The UFT/ Merck Ionic Liquids Biological Effects Database, www.il-eco.uft.uni-bremen.de
3. NIST, Ionic Liquids Database Standard Reference Database #147, <ilthermo.bouder.-nist.gov /ILThermo/mainmenu.uix>
4. R. Todeschini and V.Consonni: "Molecular Descriptors for Chemoinformatics", (2 volumes), WILEY-VCH,Weinheim (Germany) 2009.
5. Chun Wei Yap, PaDEL-Descriptor: An Open Source Software to Calculate Molecular Descriptors and Fingerprints, J. Comp. Chem., (2010):**32**: 1466-1474.

6. J.-L. Faulon, D.P. Visco Jr., R.S. Pophale, The signature molecular descriptor. 1. Extended valence sequences vs. topological indices in QSAR and QSPR studies, J. Chem. Inf. Comput. Sci. **43** (2003) 707–720.
7. J.-L. Faulon, C.J. Churchwell, J.D.P. Visco, The signature molecular descriptor. 2. Enumerating molecules from their extended valence sequence, J. Chem. Inf. Comput. Sci. **43** (2003) 721–734.
8. L. Breiman, J. H. Friedman, R. A. Olshen, C.J. Stone, „Classification and Regression Trees", CRC, Boca Raton, 1998.
9. T. Hastie, R. Tibshirani, J. Friedman, The Elements of Statistical Learning, Data Mining, Inference and Prediction, Springer Verlag, Second Edition, New York, 2009.
10. T. Chen, T. He and M. Benesty (2016). xgboost: Extreme Gradient Boosting. R package version 0.4-3. https://CRAN.R-project.org/package=xgboost
11. R Development Core Team, , R: A language and environment for statistical computing. R, Vienna, Austria,www.R-project.org 2015.

444791 Dissolution and Precipitation of Minerals Under Different Alkali Concentrations in a Reactive CO_2 Batch System

Abstract:

The primary focus of this research is to study the dissolution and precipitation of minerals and salts from Berea sandstone and brine samples taken from the Great Salt Lake in a reactive CO_2 batch system. Various experiments were constructed to measure dissolution and precipitation reaction rates under different alkali concentrations. Changing the alkali concentrations of the Great Salt Lake brine samples produced changes in dissolution and precipitation of minerals from the Berea sandstone and the Great Salt Lake brine samples. Data from ICP-MS was used to quantify mineralogical changes in the samples and Micro-CT was used to measure the porosity change in the Berea sandstone samples. The results from these analytical instruments showed patterns of mineralogical dissolution and precipitation in the post-experimental fluid that correlate to the change in alkali concentrations.

444825 Significance of Energy Intensity and GHG Emission Reduction Towards Sustainability of a Petrochemical Plant

Abstract:

Sustainability is an important pillar in today's petrochemical industry for improving process efficacy by reducing specific energy consumption, greenhouse gas (GHG) emissions, and wastewater. SABIC (Saudi Basic Industries Corporation) like other giant petrochemical companies has set stringent sustainability targets and investing huge resources to transform its plants from the present position to the first quartile in each category.

SABIC is the world largest producer of ethylene oxide and ethylene glycol. For each ton of glycol produced, 0.2 to 0.5 tons of CO_2 is emitted as GHG. The significance of sustainability "in general" while the methodology adopted to reduce energy intensity, carbon footprint, and wastewater "in particular", will be discussed thoroughly in our talk.

444827 Syngas to Ethylene Glycol: A Technology Overview

Abstract:

SABIC is the world largest MEG producer petrochemical company and most of its MEG is being produced through the hydrolysis of EO. Ethylene oxide (EO) is a simplest form of epoxides and due to its molecular structure, acts as a versatile intermediate for a variety of chemicals such as; monoethylene glycol (MEG), di- and triethylene glycols, ethoxylates, ethanol amines, and glycol ethers. The growing demand of MEG owing to its downstream utilization in polyester industry has prompted the research of an efficient, cost effective, and environmentally benign EO production process.

The major feedstock for MEG are ethylene and syngas. The heavy dependence of MEG producers on ethylene and its scarcity in some parts of the world promoted the need of alternate feedstock for highly demanded MEG. The technological developments utilizing syngas as a feedstock will be discussed along with the commercial analysis and future growth in EO/EG business.

444834 Stability and Rheology of Emulsified Acid in Waste Oil: Effect of Emulsifier Concentration, Mixing Speed, Temperature and Droplet Size

Abstract:

Acid in oil emulsion is used in the industry for stimulation purposes. In particular, HCl as the acid phase and diesel as the oil. Emulsifying HCl has numerous benefits over regular HCl. Perhaps the main reason is due to deeper penetration. Apart from that, less corrosion damages are caused since the external phase is hydrocarbon. Several studies showed the success of replacing diesel as an external phase with other hydrocarbon oil, such as crude oil and xylene.

This work utilizes the extra hydrocarbon left unused –or sometimes dumped- from refineries. This unused hydrocarbon is referred to as waste oil. First the chemical composition of waste oil was studied using HPLC, SARA fractionation as well as FTIR. This helped in selecting the proper emulsifier agent for the waste oil. The HCl-waste-oil emulsion was prepared using 15wt% HCl with a ratio of 70:30 of acid-to-oil. Extensive work was done to ensure reproducibility of data. This study was done to experimentally investigate the effects of distinct important variables on thermal stability and apparent viscosity of the HCl-waste-oil emulsion. These variables are; emulsifier concentration, mixing speed and the addition of corrosion inhibitor. All experiments were conducted at a high temperature 120˚C. To assess the quality of the HCl-waste-oil emulsion, droplet size was measured as a function of temperature (up to 100 ˚C).

The results showed that the HCl-waste-oil emulsion is a shear-thinning fluid. Power-law model was applicable seamlessly to all of the apparent viscosity data for all of measured temperatures. The optimum conditions were found to be; 0.5vol% and 600 rpm for emulsifier concentration and mixing speed, respectively. Apart from that, the emulsions prepared with corrosion inhibitor were less stable compared to the ones without it. However, the addition of corrosion inhibitor tended to generate monodispersed emulsion but bigger size droplets for temperatures 25˚C and 50˚C and polydispersed droplets for higher temperatures. Droplet size measurements showed an average size of 1.47 to 3.09 μm. Overall, the lab results showed a promising potential for the HCl-waste-oil emulsion to be used in the field.

This work takes into account; reducing the budget of acidizing job and at the same time environmental concerns. It makes use of the unwanted hydrocarbon from refiners. And, it saves the environment by using the waste oil instead of dumping it.

444994 Multiscale Simulation of Proteins Motion in the Microchannel

Abstract:

Protein adsorption is an important phenomenon, which plays a prominent role in chemical, industrial and medical areas. In this work, the dynamic behavior of the egg-white lysozyme has been investigated in a micro-channel with a multiscale modeling approach. The Molecular Dynamic simulation, free energy computation and Langevion Dynamics (LD) simulations have been combined to study lysozyme adsorption process on Au (111) surface. The gold surface has strong attraction to proteins. The protein-surface adsorption or desorption free energies is calculated. From the case studies of LD simulations, the surface has effects on the protein dynamic adsorption process. By the comparison of LD, MD and the experiments, the time step is concluded as an important factor to study protein adsorption behavior at the water-surface interface. Fine time step at the scale of 1 picosecond is necessary to provide enough accuracy to resolve proteins' dynamics trajectories.

445004 Grand Challenges in Future Energy and Chemicals: Climate Change and Sustainability As Drivers for New Technology and Innovation

Abstract:

Global demand for energy will double by 2050, driven by increasing population and affluence especially in emerging economies. While fossil fuels will continue to underpin the energy supply for the near future, concern over impact on climate change continues to grow, driving interest in expanded use of renewables and development of carbon capture and sequestration or re-use solutions. In addition to CO_2, water use and air emissions are critical issues to be addressed in future and current energy supply, as is energy storage for effective use of renewables. This presentation will provide an overview of what industry is doing today to address these issues, including the factors and metrics used in decision making.

445029 Electrospun Nanocomposite Fibers for the Removal of Heavy Metals in Water

Abstract:

Uniform electrospun nanofibers with small diameters were successfully prepared by systematically optimizing the electrospinning parameters including polymer concentration, salt loading, applied voltage, tip-to-collector distance, and flow rate. These parameters are shown to affect the fiber morphology and the addition of iron (III) nitrate salt was found to be able to reduce the fiber size. Due to the conductive nature of iron salt in polymer solution, even a relative lower voltage of 6 kV could generate a stable fluid jet and yield the nanofibers. The thermal properties of the composite fibers were investigated by thermal gravimetric analysis (TGA) and differential scanning calorimetry (DSC). Rheological studies were also conducted. The composite nanofibers were further carbonized to carbon nanofibers. Applications of the nanofibers in environment remediation have shown promise.

445030 Heavy Metal Detection in Treated Water Via a Fabric Nanocomposite Sensor

Abstract:

We present the conception of an electrochemical sensor for on-site/portable water analysis. This flexible sensor was developed to quantify the amount of heavy metals (in real-time), alleviating the wait time, large sample size, possible contamination, and expensive analytical equipment associated with current procedures for water analysis. This sensor is composed of a multiwalled carbon nanotube (MWNT) functionalized polymer mat, for a flexible yet conductive sensor. The MWNTs are then functionalized with a cyclo-oligomeric calixarene, which is shown to selectively form a supramolecular complex with metal ions. Upon complex formation, the charge carriers are drawn away from the carbon layer and the detection is prevalent at appropriate levels. The optimization of MWNT and calixarene functionalization, as well as the selectivity and sensitivity of the sensor will be discussed.

447119 Development R&D of Manoir Industries and Emphasis on Coating

Abstract:

The Senior VP Technical Support & Projects will present the development R&D of Manoir Industries and especially the coating (it is our lattest development) as well the services we can provide such as metallurgy expertise, advice, convector cleaning, site welding, welding supervision in order to increase the furnace productivity. So it is 60% R&D and 40% services. Manoir Industries manufactures cast components for the petrochemical market and develops high temperature alloys such as the Manaurites.

444478 Biogas Reforming with Addition of Carbon Dioxide to Syngas over Ni-Based Bimodal Pore Catalyst

Abstract:

Methane dry reforming of (MDR) has been widely studied due to its important applications in producing of syngas ($CO + H_2$), removal of two greenhouse gases, and upgrading of biogas (mainly composed of CH_2 and CO_2) into value added chemicals. Compared to the noble metal catalysts, nickel-based catalysts are suitable for industrial scaled-up MDR due to their high activity and low cost. However, nickel-based catalysts have been encountering a significant challenge that is the active metal sintering and coke deposition, leading to the catalyst deactivation and then the poor stability. Nevertheless, it has been reported that the deactivation of nickel catalysts can be suppressed by adding promoters like strong Lewis bases with the enhancement of chemisorb CO_2, such as MgO and CaO, and like lanthanide elements with the capacity of oxygen storage and release, such as CeO_2 and La_2O_3. Al_2O_3 is a commonly applied catalyst carrier own to its good pore size dispersion, high specific surface area and high mechanic strength.

Herein, we report that the bimodal NiCeMgAl catalysts with different NiO-loading were synthesized by the refluxed co-precipitation method and were evaluated for MDR performance by using real biogas from anaerobic digestion of food wastes. The sample containing 15wt% NiO-loading (Ni_{15}CeMgAl) was found to be active enough at 750 °C with a high CH_4 conversion of around 90%. Ni_{15}CeMgAl catalysts kept their bimodal porosity after reduction and MDR process. The evolution of the Ni_{15}CeMgAl catalyst before and after MDR process was investigated by SEM, TEM, XRD, and TGA techniques. The correlation between the structure evolution and catalytic performance change was also discussed.

444551 Continuously Adjustable, Molecular-Sieving "Gate" on 5A Zeolite for Distinguishing Small Organic Molecules By Size

Abstract:

Zeolites/molecular sieves with uniform, molecular-sized pores are important for many adsorption-based separation processes. Pore size gaps, however, exist in the current zeolite family. This leads to a great challenge of separating molecules with size differences at ~0.01 nm level. Here, we report a novel concept, pore misalignment, to form a continuously adjustable, molecular-sieving "gate" at the 5A zeolite pore entrance without sacrificing the internal capacity. Misalignment of the micropores of the alumina coating with the 5A zeolite pores was related with and facilely adjusted by the coating thickness. For the first time, organic molecules with sub-0.01 nm size differences were effectively distinguished via appropriate misalignment. This novel concept may have great potential to fill the pore size gaps of the zeolite family and realize size-selective adsorption separation.

16AIChE
Spring Meeting, Houston, TX

The Principle of Intrinsic Reality
Solving the World's Woes through Chemical Engineering and Economics

Jed Moyano Bellen, SCPM, PMP
Hyundai Engineering & Construction Co. Ltd.
9 Temasek Blvd., #27-02 Suntec Tower 2, Singapore 038989
jed.bellen@hdec.co.kr

Prepared for Presentation at
American Institute of Chemical Engineers
2016 Spring Meeting
11th Global Congress on Process Safety
Houston, Texas
April 10-14, 2016

UNPUBLISHED

The Principle of Intrinsic Reality
Solving the World's Woes through Chemical Engineering and Economics

Jed Moyano Bellen, SCPM, PMP
Hyundai Engineering & Construction Co. Ltd.
9 Temasek Blvd., #27-02 Suntec Tower 2, Singapore 038989
jed.bellen@hdec.co.kr

Keywords: Principle of Intrinsic Reality, Chemical Engineering, Economics, Sustainability

Abstract

Human beings in profound rationality always seek for explanations of the phenomena around them. To understand these phenomena, human beings have developed systematized body of knowledge as a practical way of dealing with life's challenges. Among the systematized body of knowledge developed are chemical engineering and economics.

An avant-garde way of looking at each of these disciplines leads to the Principle of Intrinsic Reality. This principle relates the fundamental laws of the two disciplines to glean insights on the constraints imposed by conservation laws on economic systems at both the macro and micro levels.

It can be argued that if mass and energy in the universe are constant, then logically diminishing returns and marginal utility are bound to be observed. This logic is quite evident in a microorganism culture which follows the Monod growth model.

It can also be thought that conservation laws imply scarcity of resources. As such, it enjoins everyone to find optimal and efficient use of resources in the context of trade-offs, production decisions and sustainability. It even calls for the development of new technologies to produce more outputs from less inputs (i.e. shifting of production possibility frontier to the right). In this way, the Principle of Intrinsic Reality leads to the imperative for sustainable optimization.

Furthermore, it is deemed that the conservation laws can explain the phenomenon called "polarity of wealth." As mass and energy when harnessed create wealth, then the constancy of mass and energy implies boundary for wealth creation. In effect, wealth can neither be created nor destroyed but can only shift from one entity to another thereby creating polarity in wealth distribution. This assertion can be the key towards more equitable restructuring of economic systems to close income gap and thereby result to stable and resilient populace.

In conclusion, this paper envisions for a paradigm shift from traditional money-centric view of chemical engineering and economics to a synergistic convergence of disciplines which can powerfully tackle the world's woes and thereby effect sustainability.

1. The World in which We Live Today

The past few years saw unprecedented global challenges which have put into question existing economic models and scientific paradigms. Among these challenges are the dystopian global economy, climate change, environmental degradation and terrorism. The magnitude of these challenges has grown to a profound level that newer ways of thinking and perspectives become necessary. With that end in mind, this paper attempts to look at the fundamental laws of chemical engineering and economics in new ways to recognize our intrinsic reality from which these pressing challenges can be tackled at large.

2. The Principle of Intrinsic Reality

It is an observable fact that advancement in science increasingly blurs boundaries of various disciplines in tackling proactively the pressing global issues or problems. Within this purview, two seemingly diverging disciplines namely chemical engineering and economics can be taken together to glean insights on our intrinsic reality and what it means for our economic and ecological systems on a larger scale.

In the author's contention, intrinsic reality is grounded on the fundamental laws of chemical engineering and economics. These laws are the Conservation Laws (Chemical Engineering) and Law of Diminishing Returns (Economics). It can be argued that the Law of Diminishing Returns is observable because of the constraints imposed by the Conservation Laws. As such, this leads to the Principle of Intrinsic Reality which can be stated thus;

> *In any system, both at the macro- and micro-levels, the limiting factor for growth is the constancy of potential inputs.*

Potential inputs mean not the actual inputs to processes but the constant material or energy available which can be harnessed to create wealth. The reasoning or proof of this principle is presented in subsequent discussions.

2.1 Proof Based on the Monod Growth Model

The best illustration to understand the Principle of Intrinsic Reality is the Monod Growth Model. Monod in his 1949 seminal paper presented an empirical relation between exponential growth rate and concentration of essential nutrient. The data was fitted into hyperbolic equation:

$$R = R_K\left(\frac{C}{C_1 + C}\right)$$

In this equation, R is the growth rate during the exponential phase; R_K is the rate limit for increasing concentrations of C; C is the concentration of the nutrient; and C_1 is the concentration of nutrient at which the rate is half the maximum.

Figure 1: Graphical Representation of Monod Growth Model

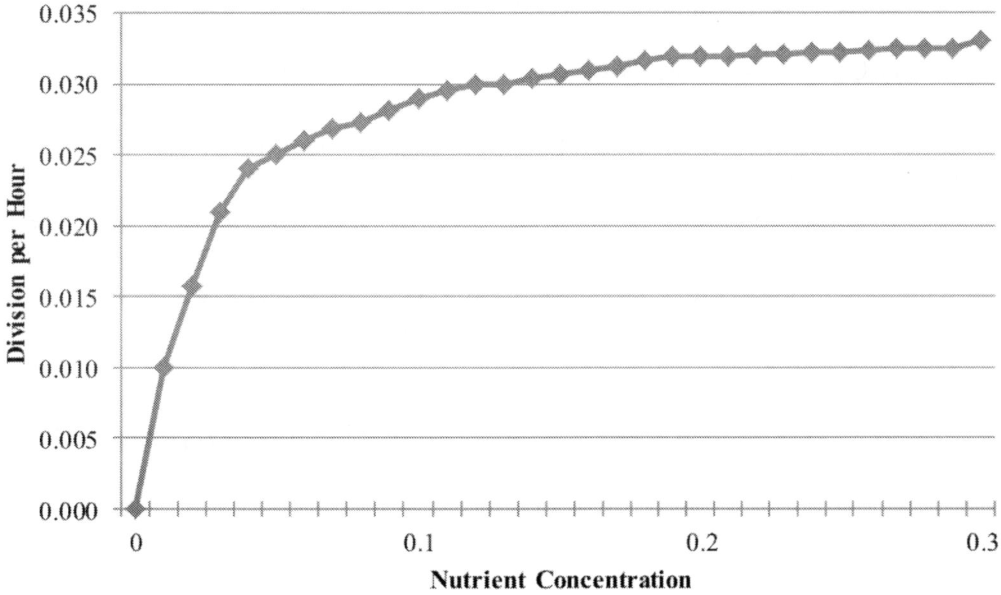

Monod (1949) further pointed out that the limiting factors commonly found are exhaustion of nutrients, accumulation of toxic metabolic products and changes in ion equilibrium.

The foregoing discussion has been observed at the micro-level with the bacterial culture. At the macro-level (i.e. the world at large), analogous phenomena are also observed to occur because of the same constraint imposed by the constant quantity of resources (i.e. material and energy). One example of these phenomena is that human beings cannot grow indefinitely. If they do, then how can the world sustain that growth with limited available resources (e.g. foods) to support it?

Another analogous phenomenon is in population growth. Thomas Malthus (1789) predicted that when population growth exceeds agricultural production, the population level is forced to return to subsistence-level conditions. This phenomenon as Malthus (1789) believed is due to the restriction of available resources supporting population growth, which others have claimed to be exploding as the earth becomes over populated.

In a surprising manner, a similar phenomenon can be observed in production processes in which the increase in a single factor of production assuming all other factors constant, results not to more marginal output but to a reduced one. This was attributed by classical economists, Thomas Malthus and David Ricardo, to the decreasing quality of inputs while neoclassical economists attributed this to the disruption in the entire production process as additional units of input are added to a fixed amount of capital. But in view of the Principle of Intrinsic Reality, the phenomenon is viewed intrinsic to the production system in which case infinite amounts of mass or energy cannot be processed with perfect efficiency. Hence, the law of conservation of mass and energy comes into the fore.

It can be deduced from the discussion that there is an underpinning intrinsic reality to the more obvious phenomena in the world from the micro-scale (bacterial growth) to the macro-scale (human growth and production processes).

2.2 Proof Based on Diminishing Marginal Utility

Another way to prove the Principle of Intrinsic Reality is through the diminishing marginal utility observable among humans and perhaps other organisms as well. Marginal utility, as defined in economics, is the gain from an increase or the loss from a decrease in the consumption of goods or services. The diminishing marginal utility is an observable phenomenon in which the utility or satisfaction an individual gains diminishes as he or she consumes more of the same unit of goods or services. Graphically, this can be represented as in Figure 2.

Figure 2: Graphical Representation of Diminishing Marginal Utility

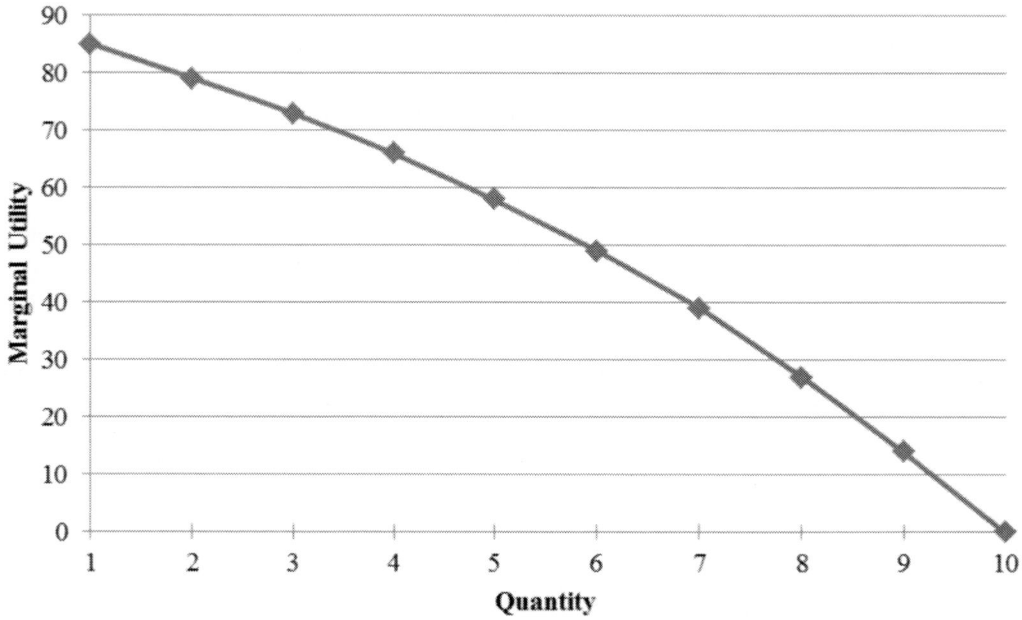

It can be deduced from this phenomenon and in support of the discussion in section 2.1 that human beings cannot grow indefinitely due to the fact that in our intrinsic reality, there is always a limit so as the constancy of available resources (i.e. mass and energy) can support our existence.

This means that in view of our intrinsic reality, we cannot have infinite greed otherwise an upsurge of unintended consequences are bound to happen (i.e. environmental degradation, terrorism and eventually demise of humanity).

Therefore, recognizing our intrinsic reality is the first step towards solving the world's woes and thereby resulting to sustainability and security.

2.3 Expanding the Production Possibility Curve through R&D

The caveat here is that the constant mass or energy (i.e. the limits) leads to trade-offs. Trade-offs is encountered when personal or business decisions are made. In making these decisions, one has to ensure that he or she becomes better off with his or her chosen option. We also realize in our life that we cannot get all that we want and we have to make trade-offs so long as it places us in a better or preferred position. This is more of a philosophical discourse, but it goes to show that the concepts of chemical engineering can be as useful in life and in business as it is in chemical processing.

To tackle these trade-offs, the Production Possibility Frontier in economics can be used. In view of the Principle of Intrinsic Reality where growth is limited by the fixed amount of material and energy, the Production Possibility Frontier needs to shift to the right as shown in Figure 3.

Figure 3. Graphical Representation of Expanding the Production Possibility Curve

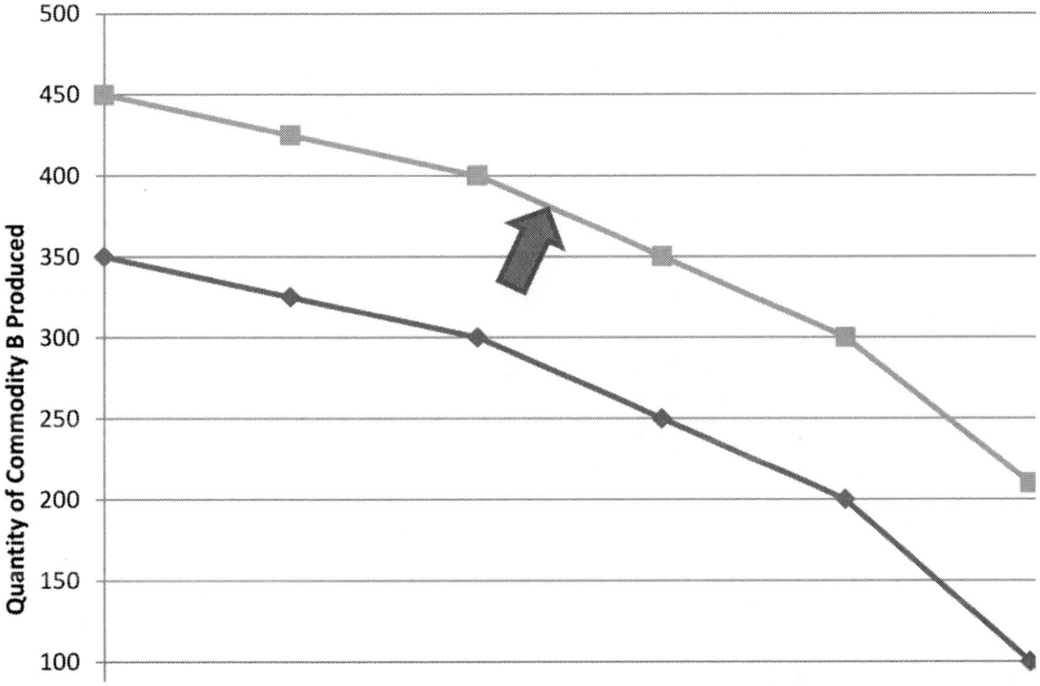

This shift has an implication on the practice of chemical engineering as Research and Development (R&D) is needed to achieve the said shift. R&D enables the economy to produce more of the two commodities (i.e. Commodity A and Commodity B) resulting in economic growth. The key challenge here to Chemical Engineering, particularly on Chemical Engineering R&D, is to look at the bigger picture where economic growth emanating from R&D output is pivoted on sustainability; considering its impact to the environment and the society at large.

3. Paradigm Shift in the Practice of Chemical Engineering

"Money" or "Profit Maximization" has been at the center of the practice of chemical engineering (Smith 2005). Various mathematical optimization methods are used to design systems, processes or components of processes with the end in mind of optimizing profits. This mental model however, will no longer work in our situation today as we face economic, ecological, social and spiritual crisis (Scharmer 2010).

To address this, Scharmer (2010) pointed out seven acupuncture points to shifting capitalism into a regenerative ecosystem economy. Within this purview, the practice of chemical engineering has a greater role to assume. For example, Scharmer (2010) suggested that the design of all production and consumption cycle be completely earth-to-earth, i.e. without necessitating for landfill and that it has to co-evolve with the natural ecosystem. In this regards, the way we do optimization should be revamped into sustainable optimization. By sustainable optimization, it means optimization of the entire system and adding in environmental as well as societal objectives to the equation.

In this definition, two things need to be considered. The first one is systems thinking which emphasizes the understanding of the entire system so the interaction among components will be known and unintended consequences mitigated (Senge 2006). This is a holistic approach to problem solving and finds enormous use in process design.

It should be noted however that the construction of processing plant, whether it is a greenfield or a brownfield project, is rooted on projected market demand. In some instances, economic volatility necessitates companies to experiment with multiple pathways to survive. In this circumstance and within the purview of systems thinking, design thinking comes into the fore. Design thinking is a collective principle emphasizing empathy for users, doing prototyping and tolerance for failure (Kolko 2015). This principle finds enormous use in chemical engineering designs such as in product development, process development, systems improvement and the like. This is also useful for simplifying or humanizing chemical engineering designs with the end in mind of contributing to sustainability. It also has to be noted that of paramount importance in using this tool is imagining the future with the view in mind of status quo and the likely outcome for sustainable development.

The second consideration is the inclusion of environmental and societal objective functions to the objective functions (economic criteria) of profit, cost, energy and yield (Edgar, et al. 2001) in optimization. The environmental objectives can be the impact or risk to environment and the societal objectives can be the impact or risk to society or common good.

These are just few examples of the needed paradigm shift in chemical engineering practice in consideration of the principle of intrinsic reality. But there can be more. A picture of the envisioned new paradigm for the practice of chemical engineering is shown in Figure 4.

Figure 4. New Paradigm in the Practice of Chemical Engineering

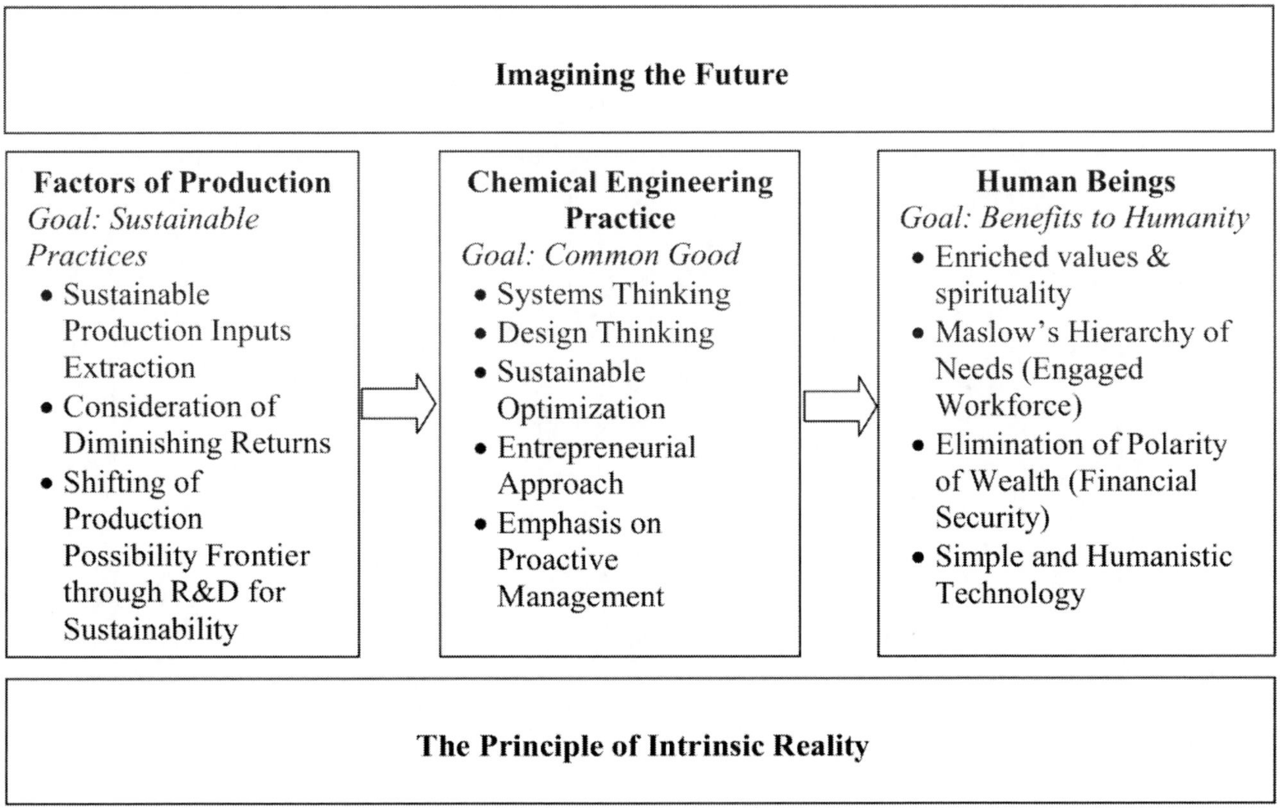

It should be realized that at the core of our economic systems is chemical engineering which can drive the economy into positive direction with the least unintended consequences. Related to that is capitalism in which upgrading of its operating systems can affect the practice of chemical engineering to a substantial degree. A discussion of this is provided in the following section.

4. Wealth Creation and Its Imperative for Sustainability

The development of this paper was inspired by the book "Freakonomics" in which the authors Steven D. Levitt and Steven J. Dubner (2005) asserted that economics can explain almost everything. It came to mind that if economics as a basic science can explain almost everything, then why not chemical engineering which was founded on chemistry, the central science?

Reflecting on it deeper, while explanations clear our doubts, we need more than that so we can tackle the world's woes. As such, this author thought that perhaps combining the two disciplines of chemical engineering and economics can be a potent force towards this end. How this can possibly be done is discussed in this section.

The starting point here is a review of the economic history of the world. In his book "A Farewell to Alms", Gregory Clark (2007) reviewed the economic history of the world in which he divided it into two periods; the period before 1800 and the period after. This is graphically represented in Figure 5.

Figure 5. Shares of Major Economies in World GDP through Time

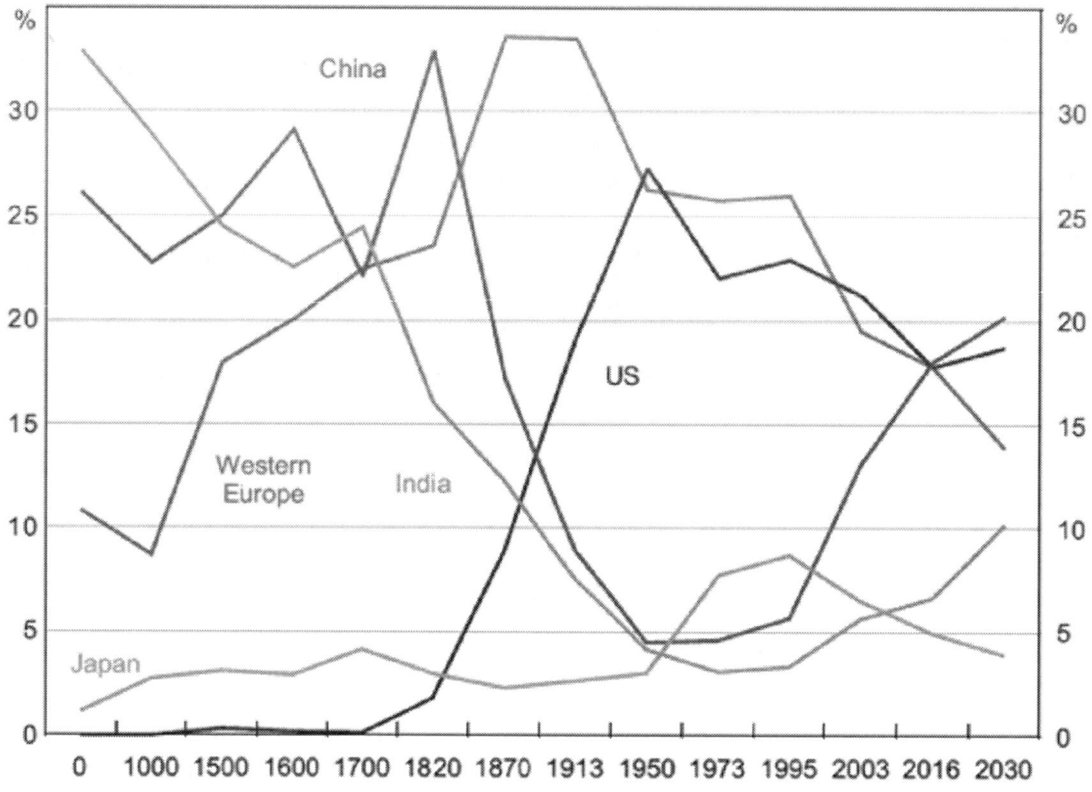

Note: Based on purchasing power parity (PPP) valuation of country GDP

Sources: International Monetary Fund (IMF), *World Economic Outlook* database, April 2011; Maddison (2006)

Thompson (2012) interpreted this graph by stating that "everything to the left of 1800 is an approximation of population distribution around the world and everything to the right of 1800 is a demonstration of productivity divergences around the world." In other words, everything to the left before Industrial Revolution depicts economic power by the sheer size of population and everything to the right after the Industrial Revolution shows the effect of mass production on country's economic output.

Clark (2007) on the other hand called the era before the industrial revolution, i.e. ca. 1800, as Malthusian era in which income per person varied across societies and epochs but growth is not observed. As Clark (2007) explained, this was due to a mechanism called Malthusian trap in which short-term increase in income through advancement in technology is dragged down by population growth. He further explicitly stated:

> "Economics as a discipline arose in the dying decades of the Malthusian era. Classical economics was a brilliantly successful description of this world. But the torrent of goods unleashed by the Industrial Revolution not only created extremes of wealth and poverty across nations, it also undermined the ability of economic theory to explain these differences."

In view of this statement, Figure 5 is reinterpreted in this paper. The contention of this author is that *polarity of wealth* can be gleaned from the graph. *Polarity of wealth* means the uneven concentration of wealth in the global economic system as some parts of the world are richer than others. The polarity can also be observed to shift as some richer parts of the world decline in economic power and some poorer parts of the world rise in economic wealth.

A historical example can be gleaned from Figure 5. India and China up until 1500 CE had 50 and 60 percent share of the world's economy, respectively. In 1 CE, China had a population of almost 60 million, while the area of the present day United States had a population of 680,000. In the late 18th century, the Industrial Revolution rendered countries with large population, just countries with the largest populations.

Looking at North America, it took them 1800 years to overtake China's economic output. However, by 1950, the US economic output was three times as great as China, even though its population is three times less. Comparing China with UK in 1913, the UK had a population of 45 million and China had 437 million but the economic output was almost the same. Analyzing what happened in Europe and later in the United States, Europe was on the lead even before 1800 (Nye 2012).

In 2010 according to data, China overtook Japan as the second biggest economy, though the numbers may not be strictly comparable. Japan's data are seasonally adjusted while those for China were not. Angus Maddison suggested that China and India were the biggest economies in the world for most of the past 2000 years. That is why economists find it a mystery why they fell so far behind and now they are flourishing ('China's economy' 2010). This may not be mysterious considering the *polarity of wealth* as was mentioned earlier and that this polarity can be argued to emanate from the Principle of Intrinsic Reality.

The reasoning here is that the world has constant mass and energy. Assuming that all these mass and energy can be harnessed to create wealth, it could mean that there is a limit for wealth creation. Theoretically, the limit can be expanded indefinitely through R&D but experience demonstrates that this was not achieved yet or getting anywhere near being achieved. As such, the constant mass and energy vis-à-vis the limit for wealth creation can be thought of as the reason for the polarity of wealth, i.e. wealth can only shift from one entity to another.

The implication of this assertion can be demonstrated by the fact that the 1% richest have accumulated more wealth than the rest of the world combined (Oxfam International 2016). The world has never seen 100% of its population to be wealthy. As such, time and again, equitable distribution of wealth is called for and mechanisms to achieve such should be made a priority in government policies.

In a more concrete way, Scharmer (2010) suggested measures to address this growing inequality which is consistent with the mindset espoused by the Principle of Intrinsic Reality. These measures are as follows:

- Creation of a New Class of Common Property Rights

The creation of common property rights is intended to address the cost to society of private economic undertakings. For example, a private company is extracting natural resources from a certain place. While the private company can profit from that endeavor, the community in that place might suffer the undesirable consequences (e.g. flooding due to forest denudation). To ensure sustainability and institutionalize the voice of future generations in this case, a trust can be established which will own the rights and distribute funds (e.g. from cap-and-trade system for carbon emissions) to citizens (one share per person), communities and the government for investment in public infrastructure.

The trust is an effective means because it does not function like a company (driven by profits) or government (driven by special interest groups) but that it is accountable for long-term sustainability of the specific commons it manage for future generations.

- Creation of a Human Right to Basic Income

One way to ensure equitable distribution of wealth is to create a human right to basic income for every human being. This right when combined with free or affordable access to healthcare and education would create a playing field where everyone has a chance to pursue aspirations and dreams. In other words, it enables individuals to put their real creativity for the benefit of larger community. Another intention of this measure is to free up paid labor (a form of dependence) so that human beings can enhance their creativity and entrepreneurial skills.

- Inclusive Capitalism

To eradicate the *polarity of wealth*, capitalism should be made inclusive so that money and capital flows are directed to all sectors of the economic system. Currently, our money system leads to accumulation of oversupply of money and capital in high financial but low environmental and social gain areas. At the same time, there is an undersupply of money and capital in areas that address societal and community needs. It implies that there is too much money in areas where it is not needed but not enough on areas where it is needed. To address this imbalance, it is recommended to redesign the system so that the current extraction-based financial system is replaced by institutions which can support the real economy and thereby serve the ecology and community.

It should be noted that shifting of wealth is also observed in various geographical regions. When this occurs, geo-political instability can result. This goes to show that if we want lasting stability and security within the limits of our intrinsic reality, then we should not tolerate unabated greed. Otherwise, *polarity of wealth* will continue to persist and we cannot realize our full potential as individuals and collaborators for common good.

5. Way Forward…

Recognizing our intrinsic reality leads us to greater understanding of our very nature bounded by the constancy of mass and energy. It shed light on why growth stabilizes after addition of more inputs. Within this purview, trade-offs are also encountered which necessitates for R&D. R&D initiatives however should be underpinned on sustainability.

Looking at the practice of chemical engineering from the lens of intrinsic reality, the approach should be futuristic while considering relevant aspects related to factors of production and human beings.

The bigger picture shows that Engineers particularly Chemical Engineers are major players in driving the real economy. As Scharmer (2010) aptly puts it, "The essence of technology and knowledge creation, and the ultimate driving force of the economy, is the same: it is the human capacity to create." As such, if sustainability is to be achieved, Chemical Engineers have a greater responsibility to innovate for regenerative and social technologies. In this way, we can stem the polarity of wealth by providing affordable technology to enable disadvantaged individuals enhance their creativity, innovation and entrepreneurship. This can help them partake in the benefits of inclusive capitalism.

As demonstrated in this paper, the synergy emanating from the convergence of Chemical Engineering and Economics can powerfully tackle the world's woes. As such, there is a good reason for chemical engineers and economists alike to collaborate particularly on sustainability initiatives.

6. References

[1] 'China's economy overtakes Japan's in real terms' 2010, *The Economist*, 16 August, viewed 5 March 2016, http://www.economist.com/node/16834943

[2] Clark, G 2007, *A Farewell to Alms: A Brief Economic History of the World,* Princeton University Press, Princeton, NJ.

[3] Edgar, T, Himmelblau D & Lasdon, L 2001, *Optimization of Chemical Processes*, McGraw-Hill, New York.

[4] Huang, Y & Wang, B 2011, 'From the Asian Miracle to an Asian Century? Economic Transformation in the 2000s and Prospects for the 2010s', in H Gerard & J Kearns (eds), *Proceedings of the Reserve Bank of Australia annual conference*, Reserve Bank of Australia, Sydney, viewed 30 January 2016,

http://www.rba.gov.au/publications/confs/2011/huang-wang.html

[5] Kolko, J 2015, 'Design Thinking Comes of Age', *Harvard Business Review*, pp. 66-71, viewed 24 February 2016, https://hbr.org/2015/09/design-thinking-comes-of-age

[6] K.N.C. 2012, 'Mis-charting economic history: more 2,000 years in a single graphic', *The Economist blog*, web log post, 20 June, viewed 29 February 2016,

http://www.economist.com/blogs/graphicdetail/2012/06/mis-charting-economic-history

[7] Levitt, S & Dubner, S 2005, *Freakonomics: A Rogue Economist Explores the Hidden Side of Everything*, William Morrow, New York, NY.

[8] Malthus, T 1798, *An Essay on the Principle of Population*, J. Johnson in St. Paul's Church-yard, London.

[9] Maslow, A 1943, 'A Theory of Human Motivation', *Psychological Review*, vol. 50, pp. 370-396, viewed 24 February 2016, http://psychclassics.yorku.ca/Maslow/motivation.htm

[10] Monod, J 1949, 'The Growth of Bacterial Cultures', *Annual Review of Microbiology*, vol. 3, pp. 371-394, viewed 2 February 2016,

http://www.annualreviews.org/doi/abs/10.1146/annurev.mi.03.100149.002103

[11] Nye, J 2012, 'Fascinating new graph shows the 'economic history of the world since Jesus'', *MailOnline*, 23 June, viewed 15 November 2015,

http://www.dailymail.co.uk/news/article-2163610/Fascinating-new-graph-shows-economic-history-world-Jesus.html

[12] Oxfam International 2016, *An economy for the 1%, how privilege and power in the economy drive extreme inequality and how this can be stopped*, Oxford, viewed 5 March 2016,

https://www.oxfam.org/sites/www.oxfam.org/files/file_attachments/bp210-economy-one-percent-tax-havens-180116-en_0.pdf

[13] Samuelson, P & Nordhaus W 2009, *Economics*, McGraw-Hill Higher Education, New York, NY.

[14] Scharmer, O 2010, 'Seven Acupuncture Points for Shifting Capitalism to Create a Regenerative Ecosystem Economy', *Oxford Leadership Journal*, vol. 1, pp. 1-21, viewed 24 February 2016, http://www.oxfordleadership.com/journal/vol1_issue3/scharmer.pdf

[15] Senge, P 2006, *The Fifth Discipline: The Art & Practice of the Learning Organization*, Doubleday/Currency, New York.

[16] Smith, R 2005, *Chemical Process Design and Integration*, John Wiley & Sons Ltd., West Sussex, England.

[17] Thompson, D 2012, 'The economic history of the last 2,000 years in 1 little graph', *The Atlantic blog*, web log post, 19 June, viewed 29 February 2016, http://www.theatlantic.com/business/archive/2012/06/the-economic-history-of-the-last-2-000-years-in-1-little-graph/258676/

Acknowledgment

The author would like to thank Hyundai Engineering and Construction Co. Ltd. for the support in this endeavor. The author is also thankful to Ms. Sofia Ayu for meticulously editing this manuscript and Mr. Marcial N. Barroga for the poster design.

442612 Big Process Data Analysis Helps Improve Operational Efficiency in Ethylene Plants

Abstract:

An ethylene facility in the Middle East was able to double its furnace run lengths from its design value, while simultaneously increasing yields and time between turnarounds, without impacting tube life and without any additional capital expenditure. An ethylene facility in the USA has been achieving record production year after year. A furnace tube leak was proactively detected very early, to prevent an unforseen shutdown at another ethylene facility in the USA. A facility in the Far East was able to increase feed to a furnace capacity constrained plant by 10% while maintaining the optimum cracking severity and obtaining the best yields of ethylene.

Ethylene plants today are by and large well instrumented and have robust advance control systems in place. So much so that the volume of available data, the velocity at which this data is collected and the sheer variety of data available from the plant and the supply chain, makes it impossible for unaided operating personnel to assimilate and get his/her mind around this Data overload. As a result more than 80% of the data collected is just archived and never used. Plants end up being run on experience based intuition, a minimal level of data or software driven analysis undertaken by limited resources with limited time and old Standard Operating Procedures prescribed by the process licensor. This often results in anomalies going undetected, potential reliability problems and missing opportunities for improvement.

A unique model driven methodology for big data analysis that allows transformation of this Data available from a typical ethylene operation, into actionable information was the key to the results observed by the ethylene producers in the case examples being quoted here. This methodology allows operating personnel to drive operations based on data-enhanced insights made readily available to them. It enables operations personnel and managers with the ability to uncover truths and insights that aren't readily obvious or don't allign with conventional wisdom or intuition. The analysis from this methodology allows predictions and provides better prescriptive insights to better determine the best course of action, while continuously growing the experience "database" of managers and ethylene operating personnel at a faster pace over time. The methology allows the improvements to be tracked, quantified and controlled, so that it can be sustained and improved continuously.

This paper covers ethylene producer cases where this methodolgy has been implemented and provides a description of the methodology.

444908 Big Data Based Models of Operators Cognition and Situational Awareness during Process Operations

Abstract:

Process industries continue to suffer from accidents despite significant regulatory intervention since the mid-1980s and major developments in process control, monitoring and supervision technologies over the last four decades. Human error is widely agreed to be the major contributor to most accidents today. The traditional approach has been to look at human error as the initiating event of incidents, one that has a given likelihood of occurrence, similar to the way that a piece of hardware is expected to fail at some frequency. Just like equipment monitoring relies on detecting prognostic signatures of its impending failure, well in advance of the actual occurrence so as to trigger timely intervention, one can ask, is it possible to detect human error before it is committed?

Cognitive engineering is concerned with the mental processes of the human actor, such as perception, memory, reasoning, and motor response especially in the context of his/her interactions with other elements of a system. In this talk, we will provide an overview of recent developments from cognitive engineering such as eye gaze tracking and pupilometry, which offer real-time high velocity, high volume data on the cognitive performance of operators. We will demonstrate using results from recent experimental studies that these data can be used to build models of operators' situation awareness. Operator's situation awareness (SA) levels can be monitored using these models, i.e., when their SA is low, operators are highly likely to take incorrect control actions with the potential to cause incidents and accidents. We will also describe a number of other potential applications of these operator monitoring models and technologies.

444881 Common Myths and Misconceptions of Online Analyzers

Abstract:

One of the many goals of manufacturing in the petrochemical industry is to operate plants safely, profitably and reliably. To achieve these goals requires the management of many competing constraints including raw materials, energy consumption, safety and environmental impact etc.. Among the enabling technologies used to accomplish these goals are computers, models and online analyzers

Dow owns and supports more than 22,000 online analyzers world-wide. The data provided by these online analyzers are used to make important decisions necessary to operate and control our plants. We will highlight and discuss many myths and misconceptions associated with online analyzers in the petrochemical industry.

Title: Using Archived Process Data to Improve First Principle Models

Author/Presenter: Peter J. Ryan, PhD, P.E.

Author/Presenter email: peter.ryan@responsepc.com

Company: Response Process Consulting LLC

Motivation

While most continuous processes have "first principle" models to support activities such as developing mass and energy balances and preparing economic analysis' for debottlenecking projects, these models typically lack the granularity to describe Key Process Indicators (KPI's) that define product specifications. For example, product color is a specification that is difficult to include in a steady-state simulation because it is not known a priori what process variables influence color, and what is the mechanism of this interaction. A novel approach is presented that uses process models based on archived process data to find relationships between process variables and product specifications. Once these relationships are established, the first principle models are supplemented to include these new process insights.

Approach

All manufacturing sectors, continuous or batch, gather process data for archiving and analysis purposes. Often the amount of data gathered, and the quality of data, makes it difficult to use this resource effectively. Major drawbacks in the quality of the archived data include:

- missing data
- noise, poor signal-to-noise ratios
- correlation in the data
- accuracy
- precision

Latent variable methods are capable of developing process models based on this resource. Latent variable models such as Principal Component Analysis (PCA) and Partial Least Squares (PLS) reduce the dimensionality of the data space such that relationships between process variables and KPI's can be established even in the presence of correlated data. This type of machine learning can be used to examine large historical process data sets and determine what process variables influence a KPI, and what is the pattern of this interaction. This information is then added to the first principle model as a mechanistic process. If, for example, the product color is determined to be influenced on the temperature of a specific heat exchanger, a mechanistic process is added to the steady-state model that represents this relationship.

Results

The first column in the fractionation system of a TDI plant is the Dephosgenation column[1]. Simulated process data (similar to the process data stored in a data historian) of a Dephosgenation column was used to develop a PLS model (Figure 1). The quality vector used by the PLS model was the (simulated) product color specification. Outliers in the simulated process data were detected using the Hotelling T^2 and Squared Prediction Error (SPE) metrics, as well as visually inspecting the T1-T2 plot of the scores. The outliers were examined using contribution charts that showed the individual contributions of the upstream process variables on the color KPI. Both T^2 and SPE contribution charts were examined. Based on a scores contribution analysis, the results suggest that seven column process variables influence the color of the product. These column process variables are the reflux flow-rate, the top and sump pressures, and the four temperature indicators that show the temperature profile in the column. The mechanism proposed to capture this interaction was a linear combination of the seven column process variables, based on the relative average contributions of each column process variable to the value of the simulated color parameter. This color mechanism was then added to the steady-state model of Dephosgenation column as a calculation block within the steady-state model.

With the enhanced steady-state model, it is now possible to include KPI terms that are dependent on upstream process variables. The process interactions captured in the enhanced steady-state model make it possible to include impacts on product specifications in the economic analysis of debottlenecking and optimization projects. These methods can also be used to improve the granularity of dynamic batch models using a similar methodology.

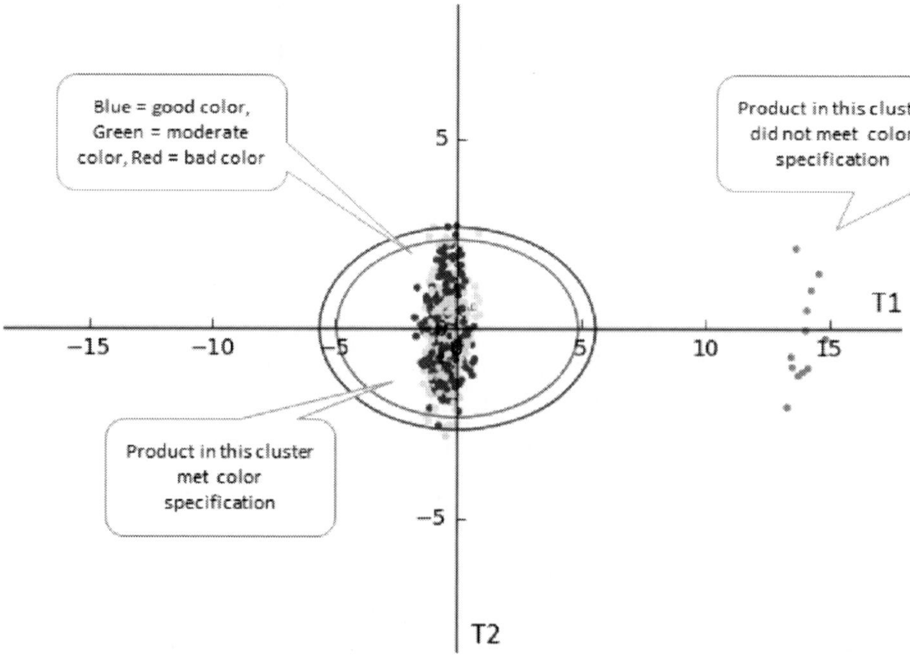

Figure 1: PLS Model developed using **continuous** simulated process data

[1] Ullmann's Encyclopedia of Industrial Chemistry, 6th Ed., 2003

444018 Integration of First-Principles Based Unit Modeling with Data Analytics Techniques

Abstract:

This paper details the infrastructure built to integrate big-data technologies in cloud with first-principle models for distributed diagnostic, predictive, and prescriptive analytics. In many industrial applications, manufactures produce process components which are geographically distributed. Oil well pumps, electric motor drives, and heavy construction equipment are examples of such products. These units require advanced and robust analytics capabilities to maximize their efficient utilization and minimize costs associated with unplanned downtime and uncertainty. While embedded control solutions manage the safe operation of the asset, analytics technologies perform fault-diagnostics, operational predictions, and performance modeling for a battery of assets. We present smart-analytics technologies to model and monitor the equipment using cloud facilities such as high throughput, multi-source data ingestion, stream processing of data, and elastic web applications. These engines include but are not limited to:

- A Proprietary model identification system which identifies the basic first-principles behaviors such as mass and energy balances, along with time-depended performance metrics such as fouling and physical performance degradation. The modeling engine is asset agnostic and provides an auto-configurable, continuously-updated unit model with high fidelity. The model retraining is automatic and is done based on the error measure produced by the engine.
- A Data clustering and prediction engine which monitors the individual unit and its trajectory using PCA and LDA analysis. This information is then used to classify the condition of the unit and provide trajectory of these conditions to the end-user. The engine is completely data-driven and uses small set of pre-labeled data to generate the model of complex operation of unit.

These engines are deployed and provisioned as cloud services using web application frameworks. Thus, these are extremely scalable and can be accessed at very high frequency. The data is ingested by cloud services and presented to the end-user via Microsoft Power-BI dash boarding service. These front-ends include standard offerings from the power bi interface as well as custom built components from Rockwell.

445007 Development of Accurate Crude Unit Model from Operating Data to Improve Refinery Profitability

Abstract:

Refinery planning models employ swing-cut representation (simplified model) of crude distillation units (CDUs), since rigorous CDU models are too large and too nonlinear for use in multiperiod planning models. Underlying assumption ("equidistance assumption") in swing cut or fractionation index models is that the crude true boiling point (TBP) curve lies in the middle between the end point of the lighter cut and the initial boiling point of the heavier cut.

This work presents a single model of a crude distillation unit (preflash, atmospheric, and vacuum towers) suitable for planning, scheduling, and RTO. The model ("hybrid model") combines first principles mass and energy balances with via partial least squares model of product TBP curves base on the feed TBP curve and operating conditions (flows, pumparound heat duties, furnace coil outlet temperatures). This enables prediction of crude distillation on par with a rigorous distillation model, with 0.5% RMSE over a wide range of conditions. Associated properties (e.g. gravity, sulfur) are computed for each product based on its distillation curve and corresponding property distribution in the feed.

We compare production planning results based on the swing-cut models and on the hybrid CDU model. For a given crude mix, refinery plan is optimized by selecting the best operating modes for the downstream process units and the best blend recipes. CDU is modelled either via swing-cut multi-mode unit or via hybrid model which can compute the best operating conditions for a given crude mix. We present the case studies for the swing-cut model with the bias terms added to the product yields and compare it to the plans based on the hybrid model of CDU.

Our case studies show that inaccurate CDU models can lead to huge reductions in refinery profits.

References:

Brooks, R.W., Van Walsem, F.D., Drury, J., 1999, Choosing cut-points to optimize product yields, Hydrocarbon Processing, **78**(11), 53-60

Castillo Castillo, P.A., Mahalec, V., 2014, Inventory pinch based multi-scale model for refinery production planning, in Klemes, J.J., Varbanov, P.S., Liew, P.Y., ESCAPE-24, Budapest

Fu, G., Sanchez, Y., Mahalec, V., 2015, Hybrid Model for Optimization of Crude Oil Distillation Units, AIChE J., DOI 10.1002/aic.15086

Guerra, O.J., Le Roux, A.C., 2011, Improvements in petroleum refinery planning: 1. Formulation of process models, Ind. Eng. Chem. Res., **50**, 13404-13418

444718 Real Time Plant Support with Focus on Enterprise Manufacturing Intelligence (EMI)

Abstract:

Dow, being one of the largest and long standing startups (118 years old), accrued immense chemical production process data. However, most of these data are underutilized and simply archived in databases. In order to extract the most significant and immediate value from that data, process design knowledge and operational experience can be used to summarize and aggregate plant data to expose the most relevant information in the right context. The goal of Enterprise Manufacturing Intelligence (EMI) is to make these high value relevant data available in real time to all levels of production management (Operations to Plant and Business Management).

Establishing new control and monitoring protocols previously unattainable due to the manner in which data systems were designed and data was stored prior to the age of "Big Data".

Furthermore, with such abundant availability of data, multivariate approaches such as chemometrics and machine learning methods can be complimentary to existing process knowledge by identifying new relationships previously "unknown". The EMI system is able to incorporate these additional process insights to further increase the breadth and performance of the monitoring framework.

Finally, to tie data trends with known and documented science to bring faster context has significantly improved the capabilities of data analytics. These improvements transcended the traditional paradigm of univariate control charting and alarm limits, and progresses towards the mastery of multivariate real-time data monitoring that is aligned to the various needs at all levels of the production facility. A robust Enterprise system and a growth strategy are the enablers of this transformation. These aspects outlined above will reinvigorate an operational focus and create a systematic memory that is robust against significant turnover in personnel and the loss of plant knowledge such turnover brings. At Dow, we want the data to start working for us, rather than us having to work to get the data.

443655 A Statistical Analysis of Sloshing Parameters for Multiphase Offshore Oil & Gas Separators

Abstract:

Liquid sloshing in multiphase offshore separators has been the subject of intense investigations for last several decades both by experiments and computer simulations. Large number scientists have worked to minimize the sloshing impacts/intensity and some others have developed new methods to describe the sloshing patterns. In addition, complex mathematical models are developed to characterize sloshing phenomenon. However, a comprehensive statistical study of the input parameters and output results is not yet been studied. In this study, statistical approach will be considered to determine the significant parameters for liquid sloshing. The factor analysis and principal component analysis techniques are considered to identify the significant parameters for liquid sloshing. Also, Partial Least Square Regression (PLSR) and Partial Component Regression (PCR) will be applied to investigate correlation among independent variables. In this regard, numerical experiments were carried out through Computation Fluid Dynamics (CFD) technique using ANSYS Fluent software. The input parameters considered here are liquid depth/tank length ratio, tank acceleration, wave frequencies, and amplitudes in various sea state conditions such as surge, sway, yaw, pitch, roll, and heave. The measured variables include hydrodynamic force, pressure, moments, turbulent kinetic energy, height of the free surface, vorticity. Mathematical correlations may be developed from the data analysis.

445013 Analysis and Mitigation of High-Impact Residential Energy Loads from Smart Meter Data

Abstract:

Residential energy demand presents a unique challenge in the overall electric grid. Although residential energy only constitutes 34% of total electricity consumption in the United States [1], the same demand disproportionately accounts for up to 50% of peak load in certain parts of the country [2]. In hotter climates, this change in load is driven primarily by air-conditioning [3]. The strain on the grid from residential energy necessitates operating additional "peaking plant" facilities for a limited period of time, which are designed for low capital cost rather than high energy efficiency and have a much higher environmental impact than their "base load" counterparts, which are operated continuously [4].

Until recently, detailed information on residential energy behaviors and patterns was limited due to insufficient data. However with the mass installation of smart meters, meters that measure and record electricity usage in high resolution, there is now a large amount of data available to analyze and predict home energy behaviors. Utilities are interested leveraging smart meter data to isolate and quantify potential changes that residential energy consumers can make in order to stabilize the grid and minimize peak loads. Although it is possible to conduct a detailed model and energy audit for an individual house, the relative benefit of one house is minimal. Instead it is most cost effective to leverage smart meter data to statistically identify energy intensive houses within a broad set of smart meter data and provide general recommendations that result in lowered peak demand

In our work we present an statistical identification method to recognize the most energy intensive houses from a broad data set of smart meter data and recommend operational changes (e.g. thermostat settings or appliance usage) or capital improvements (e.g. installation of thermal insulation or a high-efficiency appliance). An analysis of the data reveals an "energy slope" as a function of ambient temperatures and internal parameters. Household surveys and energy audit data combined with K-means clustering and parametric analysis highlight the most significant components affecting the energy slope. The analysis in turn provides an estimate of the energy reduction when considering multiple mitigation opportunities. The benefit of these suggested changes is many of the operational changes require no additional capital cost and are targeted toward minimal violations of normal preferences. The research also compares the additional benefit of higher-resolution smart meter data. A yearly data set of smart meter data from 100 houses from the Pecan Street project in Austin, TX in 1-minute time intervals provides a case study demonstrating the efficacy of the method.

[1] U.S. Energy Information Administration. Electric Power Annual Electricity Report. March 23, 2015.

[2] Wattles, ERCOT Demand Response Overview & Status Report , 2011

[3] K.X. Perez, et. al. Nonintrusive disaggregation of residential airconditioning loads from sub-hourly smart meter data. Energy and Buildings, 81:316–325, 2014.

[4] W.J. Cole, et. al. Use of model predictive control to enhance flexibility of thermal storage cooling systems. In Proceedings of American Control Conference, 2012.

444853 Big Data Analysis: A Text Mining-Based Chemical Incident Case Study

Abstract:

Chemical incidents can be mitigated and prevented by lessons learned from past incidents. Safety Taxonomy provides a useful, consistent, structured definition of safety terms, which allows better communication and knowledge sharing. Therefore, over the incident to develop an efficient Safety Taxonomy model and understanding them is very important. This paper describe safety concepts using advanced text mining methodology to conduct a comprehensive analysis, from September 1998 to October 2015 74 cases of the US Chemical Safety and Hazard Investigation Board (CSB) report and from November 1979 to April 2015 851 cases of major accident reporting system (eMARS). Text mining also allows topic clustering and analyses of Common factors and differences of each of the data. Based on the research, developing of safety Taxonomy concepts modeling and describe the topic keywords by the analysis.

Keywords : Text mining, Topic clustering, Data mining

443983 Big Data Analysis and Global Optimization for Petrochemical Planning Operations

Abstract:

Recently, the amount of collected data has experienced exponential growth, and the potential of using data for development of novel practical solutions is being realized by both academia and industry [1]. In this work, we have used high volumes of information-packed data in order to develop a data-driven planning model for an integrated refinery-petrochemical complex. Planning problems can be formulated as mixed-integer non-linear optimization models coupling process models, mixing and pooling models, mass-balance and unit connection constraints, demand and quality constraints, as well as unit and production mode selection decisions [2-3]. Models used within large-scale planning problems must maintain a balance between accuracy and complexity in order to create overall tractable optimization problems. Simplified linear formulations have been used in the past; however, they lead to suboptimal solutions due to their inability to capture the inherent nonlinear behavior of process operations and interactions.

Through this work, we aim to demonstrate the power of using past assimilated data for the prediction of future optimal operating plans, when coupling data-driven concepts, with first-principle information and deterministic global optimization. The main objectives of this work include (a) the analysis and preprocessing of noisy industrial data from 43 units present in three connected plants (one refinery and two chemical processing plants); (b) the identification of the critical correlations between input flowrate streams and input stream properties with outlet yields and outlet stream properties; (c) the identification of the form of the input-output functional forms for each of the above correlations and the global optimization of their parameters based on a least-squares minimization objective; (d) the integration of the unit operation models through connection of input-output streams; (e) the integration of pooling and blending units, demand constraints, capacity constraints and operating cost constraints; and (f) the global optimization of the complete planning model for profit maximization.

The input-output models included in this formulation range from linear, to quadratic, to polynomial, and contain parameters which are globally optimized based on the data which spans two years of operation. For automated updating of the model parameters, we have developed a user-friendly computational platform which is used to update cost, price, demand and specification parameters for the planning period of interest. The global solution of the complete planning model is found using the global optimization solver ANTIGONE [4]. Using the developed framework, we have obtained results for multiple case studies and we have obtained solutions with up to 50% improvement in profit.

References:

1. S.J. Qin, Process Data Analytics in the Era of Big Data, AIChE Journal, 60(9), 3092-3100.

2. N.K. Shah, Z. Li, M.G. Ierapetritou, Petroleum Refining Operations: Key Issues, Advances and Opportunities, Industrial Engineering & Chemistry Research, 2011, 50, 1161-1170.

3. K. Al-Qahtani and A. Elkamel, Multisite Refinery and Petrochemical Network Design: Optimal Integration and Coordination, Industrial Engineering & Chemistry Research, 2009, 48, 814-826.

4. R. Misener and C.A. Floudas, ANTIGONE: Algorithms for Continuous/ Integer Global Optimization of Nonlinear Equations, Journal of Global Optimization, 2014, 59(2-3), 503-526.

Acknowledgments: The authors would like to acknowledge the Chinese Academy of Sciences Visiting Professorship for Senior International Scientists.

444199 Detecting Changes in the Catalyst Degradation Rate of a Fixed Bed Reactor

Abstract:

Analyzing process data to understand production issues has numerous challenges. There is a large amount of data with a variety of sources of measurement variations and errors. In order to determine the cause of an undesired change in the production process, the data often needs to be analyzed for corresponding changes in the other variables. This can be especially difficult when the exact time of the change is hidden by the variability within the data set. The underlying physics can provide insights to the analysis. These physics suggest underlying transformed variables that often reduce the noise and suggest parameters that can be fitted to enforce the physics on the process. The fitting of these parameters is complicated due to the limit range of process variables that are often found in production operations.

This presentation examines the degradation of catalyst in a fixed bed reactor and explores methods to detect the event that is connected with the change in the catalyst degradation pattern of the reactor bed. This degradation results in changes in the pressure drop. For the reactors being considered, a steady increase in this pressure drop is inherent, but the rate of pressure drop growth increases for unknown reasons. The pressure drop is also dependent on flow rate and process conditions, so it can difficult to identify the actual time of the change, making it difficult to determine potential causes for the changes in the underlying degradation.

A catalyst production run can last for years, although the current work has been limited to analyzing about 60 days in each study. In these studies, the data set involves about ninety sensors with 1 minute data intervals, so over 7 million data points are being considered. There are also a variety of sources of data errors with some being random error, others being systematic or measurement drifts, and even total sensor failure. The non-random error variables can sometimes be managed by considering a physical balance equation that involves multiple sensors. Knowledge of the physics can also be used to predict other variables that can be useful in analyzing the data.

For the fixed bed reactor example, the complexity of the physics and the lack of relevant information resulted in a number of competing simplifying assumptions that could be used to build a degradation process model. Each of these potential models was considered in the data analysis in order to separate the process variability from the degradation rate changes. The comparison of these different fits and an analysis of the residue of these fits provide better identification of the time of the changes and provide insights into underlying physical changes. The analysis of changes in the data is then used to identify potential events that might be the source of the change in the degradation.

443206 An Artificial Neural Network-Computational Fluid Dynamics (ANN-CFD) Model to Predict Solid Particle Erosion

Abstract:

This study employs statistical analysis to develop an artificial neural network (ANN) along with a computational fluid dynamics (CFD)-based methodology that can develop robust and accurate erosion prediction models. Solid particle erosion plays a critical role in the design and reliability of equipment employed in the oil and gas industry. Significant erosion occurs due to solid particle loading, especially in applications involving sand production. Computational fluid dynamics has emerged as powerful tool to predict erosion in recent years. The ability to simulate multiphase flows in complex geometries using CFD makes it a valuable and less-expensive method to predict erosion. Various empirical relations have been established to predict erosion using CFD. These methods often predict erosion regions accurately, but are typically highly inaccurate in predicting an erosion rate. An order-of-magnitude error is observed in many cases.

An extensive set of experimental data was available experiments conducted on 90-degree elbow, which was used for exploratory analysis and as target data set for simulations. The current in-house erosion model is studied as a baseline. Exploratory data analysis was performed on CFD output parameters to identify influences and correlation to solid particle erosion. An artificial neural network with multilayer feed-forward model with back-propagation algorithm and Levenberg-Marquardt training was developed. This model, along with Bayesian regularization, reduced cumulative error to less than 10%, compared to more than 40% in the baseline Baker Hughes model.

443688 Stochastic Modeling Using Virtual Training Sets

Abstract:

ABSTRACT for AIChE 2016 Spring Meeting

Stochastic Modeling using Virtual Training Sets

James C Cross III

The MathWorks, Inc.

james.cross@mathworks.com

617-605-5818

In a typical process plant, instrumentation data are archived, providing a rich set of structured data that can be exploited for system identification, efficiency optimization, and anomaly detection. Combining these data with process simulation results, plant operating heuristics, and economic parameters, a detailed operating model affording the prediction of output and associated expenses can be built to guide planning decisions. This is fairly common practice.

A more complicated situation arises when an organization must decide how to operate a portfolio of assets producing the same products, for example to maximize profit over a known time horizon for a specified total output level. Examples include resource extraction from oil and gas wells, production of electricity from a pool of power generators, and water resources management. While most plants and other process equipment are designed to operate consistently at or near design capacity, market fluctuations dictate the need for part-load strategies and accordingly these types of analyses, as has happened in recent times with the downturn in oil prices.

To support planning, a single answer from a predictive model is inadequate. It can stand as a reference scenario, however diligent risk management demands that the probability of alternative answers, such as would arise from uncertainties associated with exogenous variables, be computed and compared to the reference. One could conceptually define a collection of scenarios, and use the detailed models for all of the plants in the asset pool to compute the global optimum for each scenario. While this is appealing, it is invariably impractical – the level of complexity of accurate plant models, the duration of planning horizons, and the multitude of operating constraints combine to make this an exceptionally burdensome computational challenge. Is there a way around this?

The application of machine learning methodologies to real data sets has seen tremendous growth in the past decade. In this work a machine learning approach is used to extract important relationships from the optimized operating scenarios calculated by the detailed optimization model. The simplified model that results from using these "virtual training sets" captures the most influential factors embedded in the detailed model, and can be used to run simulations across a statistically significant number of scenarios in a reasonable time.

This talk elaborates the approach and offers a framework. The methodology is motivated through use of a simple illustrative example, which reveals both the advantages and limitations of the approach. Select open questions and opportunities for further work are highlighted.

445134 ProcessMD: Capturing the Heartbeat of our Plants through Advanced Analytics

Abstract:

The Industrial Gases market is characterized by production plants that balance both capital and operating costs along with the important need for the continual supply of product. It is therefore critical to maximize asset productivity while concurrently maintaining baseline reliability requirements. In response to these operational objectives, companies like Air Products and Chemicals, Inc. have sought to acquire the requisite process insight through advanced data modeling and analytics.

With the advent of digital control systems and data historians, manufacturing companies now have a preponderance of data; however, these high-dimensional datasets are often characterized by low signal-to-noise ratio along with a high degree of correlation. Advanced analytics in the form of state-of-the-art statistical multivariate techniques can denoise process signals, as well as capture the correlation amongst these signals, transforming huge plant datasets into actionable information.

ProcessMD, a patented web-based, predictive-monitoring and fault-diagnostic platform, has been developed to provide foresight and insight about asset performance. At the foundation of this novel technology platform are predictive models built uniquely for each plant or piece of equipment. These data-driven, auto-adaptive models provide advanced alerts of subtle abnormalities and deviations from expected behavior across a whole plant or its specific components. The predictive analytics focuses on diagnostics to enable proactive intervention – unlocking value from data.

444202 Big Data Analytics in Oil & Gas

Abstract:

Big data, and big data analytics, have existed for some time in oil and gas, particularly in the form of seismic data analyzed using high performance computing clusters. However, these days, data from other sources is increasing, and tools specifically designed to perform analysis on large diverse data sets have emerged. In oil and gas, high volume and high velocity data occur routinely, while data of great variety are the norm, and significant issues around data veracity exist as well. The emergence of big data analytics in oil and gas is permitting the industry to address challenges that could not be satisfactorily handled previously. These techniques can be applied across the full value chain, from exploration to drilling to production and processing, thru shipping and handling of supply chains and trading. This talk will examine applications of big data analytics in oil and gas as well as areas where big data analytics has not been applied but could add value.

445141 Moving the Needle through Transaction Level Analytics

Abstract:

Dow's Advanced Analytics team provides actionable insights to key business and corporate leaders using a wide range of computational methods. We are continuously striving to integrate any sources of information that can make our models more timely and accurate to improve decision making. Most attempts at understanding and predicting customer demand have either been non-analytical surveys of customer needs or have used time series methods of forecasting aggregate demand over customer segments. At Dow we have developed a predictive analysis of customer demand which brings the world of analytics to the understanding of individual customers. A crucial aspect of this modeling is the understanding of the high volume and velocity of order dynamics, such as changes to existing order prices, quantities, and order delivery dates. The value of this work extends beyond better forecasting to allow prescriptive analytics for prioritizing marketing activity and earlier detection of problems in the supply chain before they become evident.

444734 Enterprise Manufacturing Intelligence (EMI) for Batch Plant Data Systems

Abstract:

Dow collects and stores millions of data across its global manufacturing sites every day using a wide variety of data historians to monitor product and process variables during production runs. Because of the complexity of these database systems, extracting and analyzing this vast amount of data into meaningful information for the plant can be complex and time consuming, thus reducing opportunities to improve certain areas of the process. In order to extract the most significant and immediate value from that data, process design knowledge and operational experience can be used to summarize and aggregate plant data to expose the most relevant information in the right context. The goal of Enterprise Manufacturing Intelligence (EMI) is to make these high value relevant data available in real time to all levels of production management (Operations to Plant and Business Management). Therefore, plants can take advantage of this information to make improvements in raw material usage, asset utilization by reducing unnecessary process wait times and customer satisfaction through improved product quality by reducing product and process variation. This applies to both batch and continuous processes. A Dow case study will be discussed in the presentation to show the benefits of EMI in a batch process.

Through the development and advancement of new commercially available systems[1], key batch process components can be monitored and tracked for a single production run through a unique batch identifier. This software provides easy and fast segmentation of production data into batches, campaigns or other logical groupings for easier analysis and production reporting to optimize process conditions. Because of the defined start and stop time throughout the batch steps, the data associated with each batch can be evaluated either individually or from one batch to another. Historically, this has not been easy to track data for a particular parameter from one batch to another. The isolation of this data provides plants key information about the production process that can help drive continuous improvement across the batch process.

Now that the batch data has been compiled into uniquely identified production runs, plants can take advantage of EMI dashboard technology to quickly understand the performance of their process. EMI allows for continuous monitoring of batches and "flags" when an abnormal process condition has occurred; based on historical batch trends and process targets. This technology also has the capability to merge process data with product analysis to ensure all pertinent information is readily available to plant personnel. This has been extremely difficult to do in the past. This can significantly improve the analysis time and prevent any reoccurrence of incidents prior to the next batch being produced. The implementation of EMI allows for continuous monitoring of long-term batch trends to drive productivity improvements proactively and utilize Dow's assets more efficiently and effectively.

[1] Marty Moran, "Extending the Value of MES Technology into New Applications – An Industry White Paper," www.AspenTech.com, 2013

443534 Using Data Mining Techniques for Process Improvement in Polypropylene Plants

Abstract:

Determination of the most significant manufacturing process parameters can be very valuable in improving product quality and plant reliability. Given the huge amount of recorded data and complexity of chemical plants, data mining is crucial to uncover information, knowledge, and patterns from the historical data.

This presentation describes how to facilitate decision making process in a polypropylene plant by using data mining techniques such as variables screening and modeling. High liquid hydrocarbon level within the polymer product is a chronical safety issue in polypropylene plants. Since the high hydrocarbon level can often be the cause of either plant shutdowns or reducing the production rate, understanding the root cause of this problem has a significant impact of the plant reliability and economics.

Data visualization and multivariate analysis techniques are used to identify relative significances of process variables and evaluate their effects on the hydrocarbon level. As a result of this study, new operational trials are proposed and tested in the plant. The result of trials is used to study the impact of changing in significant variables on reducing hydrocarbon level.

440552 Big Data Doesnt Really Mean Big Intelligence

Abstract:

Big data is the IT phrase of the moment in almost every industry. The concept of collecting huge amounts of data to learn, manage and drive your business forwards is attractive to all managers, especially as without it you may feel that you're being left behind. But without knowing how to collect data effectively, knowing what data has value, and how to effectively and efficiently use the volume of information, Big Data can be a hindrance, not a help.

In Process Safety, we rely on data coming from the field in many instances. Collecting this data can be time consuming and costly, so making the most of the data you have is paramount, as is ensuring the data you collect is seen as a valuable asset by everyone in your organisation. But turning that data into intelligence to guide you on your path to Process Excellence is what it's really all about. Data is as valuable an asset as any other. Look after it. Nutrue it. Value it. And make sure you use it.

In this paper we examine the core principles of making data work for you to give you the best usable, actionable intelligence possible.

443388 Root Cause Investigation on Defect Batches with Advanced Statistical Techniques

Abstract:

The chemical industry often involves batch processes in which online monitoring plays an important role for batch quality assurance. Often times, process conditions are electronically measured very frequently (i.e. every minute) over the entire time-span of the processing steps for each produced batch. After multiple batches have been produced, a rather large data set of the processing conditions is available to help correlate the changing of process conditions with resulting batch quality. In addition, the trajectories of the process variable levels between good and defect batches can be compared to examine if abnormal behavior exists in defect batches.

With a real Dow polymerization process example, this presentation discusses how the rather large process data set (many processing measurements per batch) is combined with the single batch quality assessment to enable meaningful analyses. In addition, several advanced statistical techniques used for root cause investigation on defect batches including partial least squares, principal component analysis, discriminant analysis, partition, logistic regression, multivariate SPC charts and batch-level modeling will be described. By using multiple statistical techniques, we were successfully able to eliminate a variety of process variables as potential root causes for why defect batches kept occurring. This conclusion enables the business to focus on the identification of alternative potential root causes and the resulting data required for assessment for ultimately solving the problem.

443657 Batch Process Monitoring By Dynamic Time Warping and k-Means Clustering

Abstract:

In the production of membranes for gas separation, the raw material goes through a series of batch processes before it is converted into a finished product. Each of these batch processes is governed by a recipe, which often involves a combination of process control logic and operator interventions. This work presents a method for monitoring the quality of a batch process by detecting deviations from the recipe, through a combination of DTW and k-Means clustering.

Historical data (e.g., flows, temperatures, pressures) are gathered from a membrane manufacturing process, which consists of several steps of variable duration. Through a combination of interactions between researchers and subject matter experts, a set of "ideal" and "bad" recipes are identified which is used as a training set for the model. Historical operation is then compared with the ideal and bad recipes, and a unified distance metric is used to then cluster the operational data using k-Means clustering. Encouraging results are reported from practice.

444697 Optimizing an Industrial Batch Process Using the Design of Dynamic Experiments Methodology

Abstract:

The recently proposed Design of Dynamic Experiments (DoDE) methodology (Georgakis, 2013) which generalized the classical Design of Experiments (DoE) approach, is applied to an industrial batch polymerization process *in silico*. The detailed knowledge-driven model of this process, unknown to the first author, is used in simulation mode by the industrial side to perform the simulated experiments. The fist author is provided with no details about the inner characteristics of the process, besides that it is a polymerization reaction which should have no more than 1% unreacted monomer and no more than 1% of impurities in the product an the end of the batch. He is also provided the operational conditions of three batch runs, from which only the first produces acceptable product quality. *The optimization objective is to decrease the batch duration while producing acceptable quality product.* An initial set of 12 experiments is designed around the single successful recipe by considering time-dependent variations of the initial monomer feeding profile and the reactor temperature. None of these runs violated the quality constraints by more than 10%, the normal variability of the process. The simulated "experimental" results leed to an initial data-driven model, which calculates the operating conditions of a run with 6.5% reduced batch duration and a product of acceptable quality. This predication is confirmed through a follow-up *in silico* experiment by the industrial side. I second set of experiments is then designed to complement the first set and aiming to reduce the batch time by 20% while keeping the expected product quality within the desired bounds. The presentation will present details on the overall methodology and an assessment of the overall results of the two sets of experiments. It will also commend on the use of the DoDE approach in real rather than simulated processes.

GEORGAKIS, C. 2013. Design of Dynamic Experiments: A Data-Driven Methodology for the Optimization of Time-Varying Processes. *Industrial & Engineering Chemistry Research*, 52, 12369-12382.

444743 Nontraditional Experiment in Manufacturing to Evaluate Measurement Systems That Count Defective Particles

Abstract:

In Dow it is common to manufacture the same product at multiple production sites across the globe. Measurement systems of laboratories at different sites are expected to obtain similar results. One way to assess laboratory performance is to conduct an Inter-laboratory Uniformity (ILU) study. Typical ILU studies have only one factor of interest: laboratory performance. Laboratories in different locations are evaluated to see how they differ in their measurement of a single sample. In this presentation, we will share a form of ILU study where we are interested in evaluating the measurement systems used to count defective particles in a powder at three different laboratories. The number of observations in the ILU study exceeded 400; traditional ILU studies in Dow usually have fewer than 100 observations. The ILU study had 3 factors that were investigated (site, analyst, and operating condition) and many challenges that had to be overcome. The distance between laboratories prevented the use of a fully randomized design. It is infeasible to create homogeneous aliquots of the same material when the variable of interest is a count of low-level defects, yet sample handling had to be minimized to prevent adulteration of samples. In this talk, we will share how we used a split-plot design and non-traditional design elements to overcome these challenges and investigate hypotheses.

443538 Catalyst Comparison in Batch and Continuous Reactor Processes

Abstract:

Catalysts are commonly used in both batch and continuous reactor processes. To screen out the best catalysts, the product yield is monitored during the reaction period. It is challenging to compare different catalysts for those scenarios, because the yields are always time dependent in both batch and continuous reactors. In this presentation, MANOVA statistics methodology is used to analyze the data which have highly time serial correlation, and some novel methods have been developed to analyze the data which has a baseline drifting.

444348 Large Data Soft-Sensor Based on Mixture of Gaussian Process Regression Model with Lasso Criterion

Abstract:

Soft sensors are used to enhance process monitoring and control of difficult to measure process variables. Moreover, modern industrial processes exhibit multi-mode behaviour due to different demands from the market. As a result, the usual assumption about uni-modal Gaussian distribution of the data is not valid. One approach to resolve this is by clustering the input data using Gaussian mixture model. However, like other clustering algorithms, experience severe performance degradation in the high-dimensional setting. To overcome this shrinkage methods have been shown to yield more stable and interpretable models. The application of least absolute shrinkage and selection operator (LASSO) criterion to mixture of Gaussian process regression model (MGPR) is proposed in this work to handle multi-mode nonlinear process with high dimension of variable. It enables simultaneous identification of significant variables and determination of important clusters. The clustering penalty penalizes the distance of the regression coefficients between clusters and thus similar data are combined into common cluster, while the regression penalty selects significant regression variables within each cluster. Case studies presented show the capability of the proposed method and its applicability to an industrial melt index process.

444970 Statistics Pattern-Based Big Data Analytics Framework for Iot-Enabled Cybermanufacturing

Abstract:

Statistics pattern-based big data analytics framework for IoT-enabled cybermanufacturing

Jin Wang[1], Q. Peter He[2], Anthony Skjellum[3], Devarshi Shah[1], Carlos Lemus[3]

[1]Department of Chemical Engineering, Auburn University, AL 36849

[2]Department of Chemical Engineering, Tuskegee University, AL 36088

[3]Department of Computer Science and Software Engineering, Auburn University, AL 36849

With the emergence of the Industrial Internet of Things (IoT) and ever advancing computing power and expansion of wireless networking technologies, a new generation of networked, information-based technologies, data analytics, and predictive modeling are providing unprecedented embedded computing capabilities as well as access to previously unimagined potential uses of data and information. These capabilities provide possibilities for new, radically better ways of doing manufacturing. Although there are different names used to describe next generation manufacturing systems, such as cybermanufacturing and smart/advanced manufacturing, the essence of these is the application of increasingly powerful and low-cost computation and networked information-based technologies in manufacturing enterprises. There is a general consensus that factories and plants connected to the Internet are more efficient, productive and smarter than their non-connected counterparts.

Manufacturing process operation databases are massive because of the use of process operation and control computers and information systems. With ever-accelerating advancement of IoT devices and other communication and sensing devices and technologies, it is expected that the data generated from cybermanufacturing systems will grow exponentially. 4 V's are often used to characterize the essence of big data: Volume (from terabytes ($\sim10^{12}$) to zettabytes ($\sim10^{21}$)), Variety (from structured to unstructured), Velocity (from batch to online streaming), and Veracity (from well calibrated and cleansed data to less trustworthy and uncleansed data).

Big Data is arguably a major focus for the next round of the transformation of advanced manufacturing. According to research by McKinsey Global Institute and McKinsey's Business Technology Office, the analysis of large data sets will become a key basis of competitiveness, productivity growth, and innovation.

In a newly funded NSF EAGER project, we propose a novel, statistics pattern-based process data analytics framework with the aim to provide smart diagnostics and prognostics for cybermanufacturing. As part of this effort, we also propose to establish an IoT-enabled manufacturing technology testbed (MTT) to explore and establish a proof-of-concept for the proposed framework. The theoretical background of the framework, the setup and test of the MTT, as well as the preliminary results obtained will be presented and discussed.

444537 Process and Performance Modeling in Production Scale Up Challenge

Abstract:

Chemical production scale up can create performance gaps in final products due to process condition variations. To streamline production scale-up, it is important to identify the key process factors of critical performance metrics. Multi-level data processing (critical performance metrics measured at different levels/stages), multicollinearity (high correlations among process factors) and "curse of dimensionality" (too many process factors and interactions compared to number of observations) together make it a challenge in statistical modeling.

In a real Dow scale up example, we applied classical regression methods (Stepwise Regression, Partial Least Square and Generalized models) and Symbolic Regression approach (linear and nonlinear model searching algorithm) for two critical performance metrics. Multiple important process factors were identified consistently across all methods for the two metrics. In this presentation, we also demonstrate model validation using an independent test data for comparing different models. Among all the models, Ridge regression has highest model prediction accuracy with the most model complexity including all process factors and interactions. Pareto front models in Symbolic Regression approach were selected for comparison with classical models in the test data.

From practical and statistical perspectives, we selected two models with relatively high prediction accuracy and low model complexity for the two performance metrics. With this work, the business can streamline process optimization by using in-process, real-time measurements to estimate critical performance metrics. Successful implementation of these process models will accelerate the progress of launching the new products in the commercial manufacturing plant.

444988 MKS Data Analytics Solutions Umetrics Suite - A Platform for BIG Data Analytics, Multivariate Process Modeling, Monitoring and Prediction

Abstract:

MKS Data Analytics Solutions contributes in the area of BIG Data Analytics as a world leading provider of user friendly software packages for multivariate data analysis (MVDA) and design of experiments (DOE) through the Umetrics Suite of products. Software solutions are graphically driven and complete with surrounding services ranging from training to consulting, implementation and support. With 25 years of experience MKS Data Analytics Solutions is proud to offer easy-to-use robust and mature solutions as well as a highly praised customer support program. The Umetrics Suite includes the products MODDE Go, MODDE Pro, SIMCA, SIMCA-Q and SIMCA-online and application implementations span a range of industries including biopharmaceutical, petrochemical, chemical, food, semiconductor and pulp and paper.

The presentation will highlight a number of success stories of off-line and on-line applications of Umetrics Suite while emphasizing graphical power and capability of rapidly extracting information from and making sense of volume datasets originating from complex arrangements of multiple data sources. Visualization and effective utilization of software tools to fulfill objectives, create return on investment and increase the bottom line will be discussed.

Examples of applications include real-time multivariate process monitoring, open and closed loop multivariate model predictive control, prediction of critical process quality attributes, increasing process understanding, process optimization and process trouble shooting. MKS Data Analytics Solutions Umetrics Suite is utilized throughout research, development and manufacturing, both for continuous and batch processes.

444832 Modeling and Monitoring of Batch Processes in Relative Time

Abstract:

Batch processes are widely used in many industries. Typically, raw materials are combined in a suitable batch vessel before a chemical, physical or biological transformation take place, resulting in the end product. In many cases the control of the batch process is recipe driven and not adjusted to accommodate raw material variation, changes in uncontrollable factors, etc. to ensure the best possible end product quality. Real time control of a batch process saves the industry money and resource due to less rework and rejects, as well as the opportunity for real time quality assurance of the end product.

There are some solutions available for batch monitoring and control but typically they are assuming equal lengths of batches, i.e. the batch starts at the same chemical or biological time t_0 and has the same number of time points for all batches. Obviously this is resulting in problems if the batches do not meet these criteria. Alternative approaches to handle uneven batch lengths include replacing time with a maturity index or using dynamic time warping. In both these approaches complications can occur if the first measurement does not coincide with the true t_0 and the batch evolution is non-linear, which is often the case.

In this presentation, a better approach accommodating both uneven batch lengths and unknown true t_0 is proposed. The approach is based on projections in the score space so all control options used in MSPC are valid and available.

443289 Leveraging Big Data with Modeling & Self-Service Analytics for Rapid Decision Making / Capstone Technology

Abstract:

Businesses today have more data than ever to interpret. Data is arriving from multiple sources at an alarming velocity, volume and variety. The speed and ability of businesses to collect, present, analyze, interpret and decide separates the elite from the masses. How do you exploit this volume of data from so many sources without crippling decision making? It would be great to throw all of this data into a Super Computer and output the optimal solution. However, optimal manufacturing decisions require not only processing the data, but also delivering it to the right people with a tool that keeps up with the speed of thought. This presentation will focus on available commercial software that does just this. 6 keys include 1) Provide access to all manufacturing and enterprise data sources 2) Use data aggregation for speed & filtering 3) Apply calculations, statistics & logic for development of KPI's 4) Layering of asset structure for ease of analysis and segmenting of data by assets & asset types. 5) Visualization, Advanced Analytics and Modeling of all underlying data 6) Use of industry standard protocols to provide easy integration of domain specific applications.

442072 Real-Time Analytic Dashboard Visualization

Abstract:

This session will focus on effective dashboard visualization to ensure real-time analytics projects achieve success. We will highlight examples of dashboard designs along the rationale for their use, and will discuss strategies for organizing and managing projects like these. We will also explore best practices for harnessing management support and funding, along with ensuring the project achieves its goals. Finally we will address implementation tips for developing organizational habits to maximize the effectiveness of the system.

444865 Put Big Data Technology in the Hands of the Engineer

Abstract:

Why do Process Engineers still rely heavily on Excel for analysing the data that has been generated for decades? Why haven't we copied the success of retail, satnav or medical Big Data solutions in the process industry? The first big reason is that the location of our data, the historian database, is not built for these types of questions and requires specific technologies that are usually expensive and require a heavy installation process before they can be used. Secondly, these technologies are built for data scientists,not for the engineer or operator. This creates a knowledge gap and discourages flexibility. TrendMiner is a Belgian software startup with many years of research experience in Process Industry analytics. This solution is to be installed and ready to use within one working day without the need of new datacenter infrastrcuture. In this presentation you'll gain insight into the technology and the cases it has already solved to date.

445137 Data Science in Process Automation: Evaluating Economic Advantage

Abstract:

Abstract

Each day businesses make decisions and take action as best they can based on available information and in the face of uncertainty. The goal of data science and analytics can be seen in terms of improving the amount and interpretation of information such that uncertainty can be reduced and quantified and better decisions and actions may be taken.

This goal – while agreeable – often raises new questions: How do we select the right data science problem that will return value for our efforts? How do we determine the performance required of a data analytics algorithm for it to be economically beneficial? How can data science techniques complement and augment traditional engineering techniques – and avoid having machine learning algorithm return simple mass-energy balance or automation configuration rules?

Honeywell provides an integrated suite of tools to monitor and automate many aspects of plants in the process industry – including refining, oil and gas, pulp and paper, chemicals, power generation, and mining.

In this presentation we will describe an approach in which the economic costs and benefits to the business can be evaluated for given data analytic solution. This economic framework can be introduced early in the data science analysis process and has proven to be an effective tool in guiding and engaging developers and stakeholders throughout all phases of the project – setting the question, exploratory analysis, model building, communication, and decision making.

16AIChE Spring Meeting, Houston, TX

12TH GLOBAL CONGRESS ON PROCESS SAFETY

Big Data Analytics Applied to Real-World Control Systems

From Instruments to Advanced Controls

16AIChE Spring Meeting, Houston, TX

12TH GLOBAL CONGRESS ON PROCESS SAFETY

Agenda

- Real-World Data from Control Systems
- Data to Information to Action
- Analysis of a Typical Plant
- Case Study
- Practical Suggestions

Real-World Data from Control Systems

Good News! You already have the data!

- Thousands of Data Points
  - Controllers (PV, SP, OP, MODE)
- Organized (for Control)
- Somewhat Organized (for Process)
- Historized
  - With Compression?

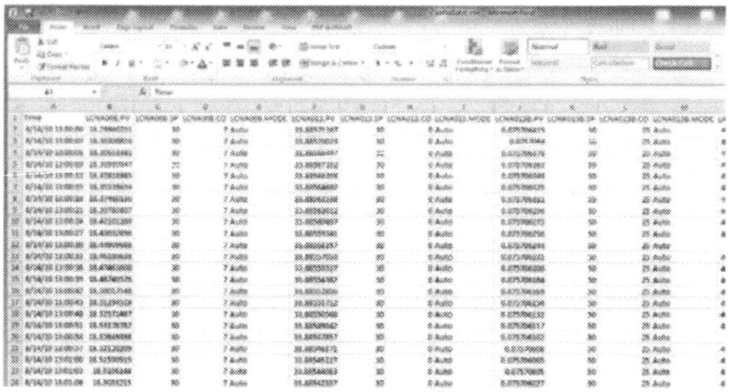

3

Data to Information to Action

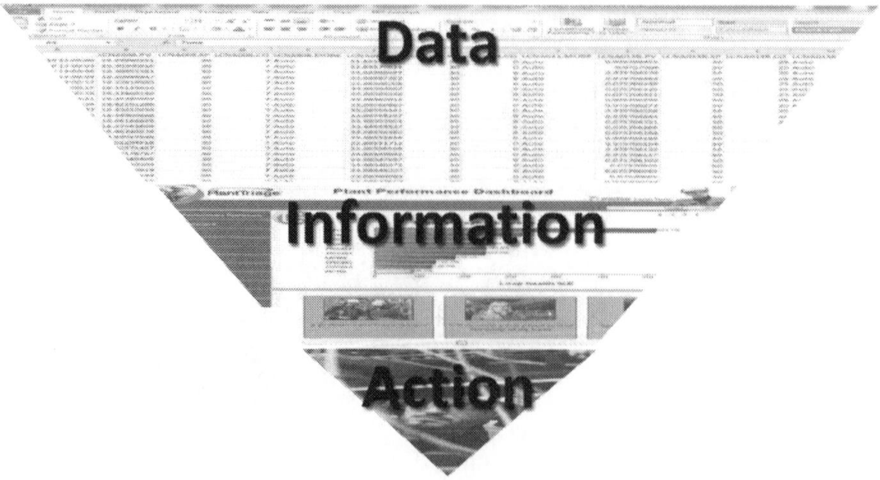

4

16AIChE Spring Meeting, Houston, TX

12TH GLOBAL CONGRESS ON PROCESS SAFETY

Data to Information

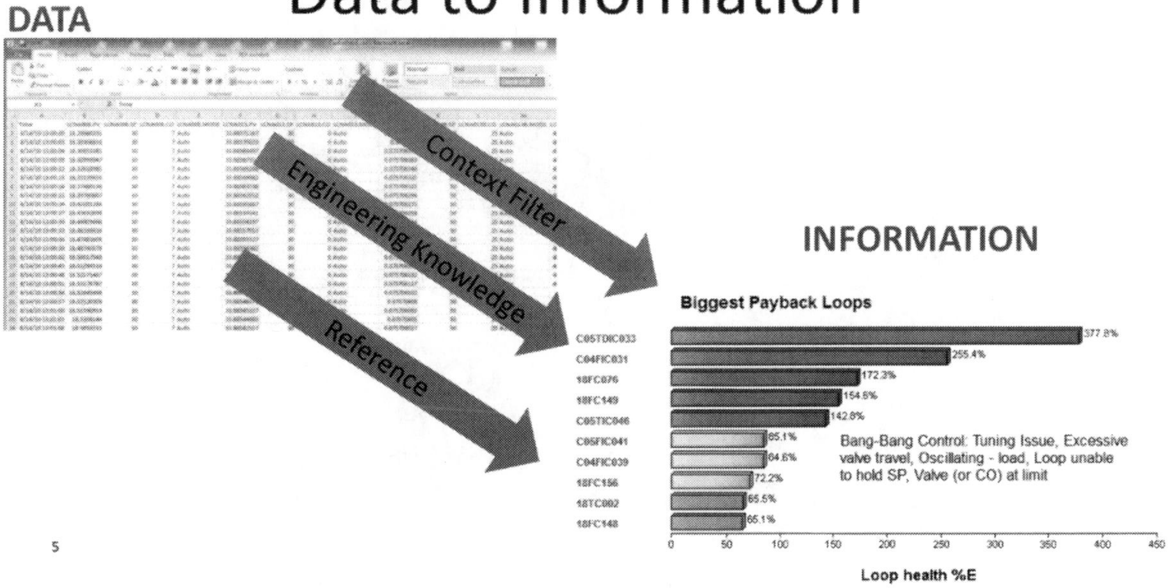

5

16AIChE Spring Meeting, Houston, TX

12TH GLOBAL CONGRESS ON PROCESS SAFETY

Simple, Yet Powerful Information from Data

Control Performance Statistics

- % of Time Loop in Manual
- % of Time Valve at Limits
- Statistics
  - Min / Max
  - Avg
  - Std. Deviation
- PV max = PV min

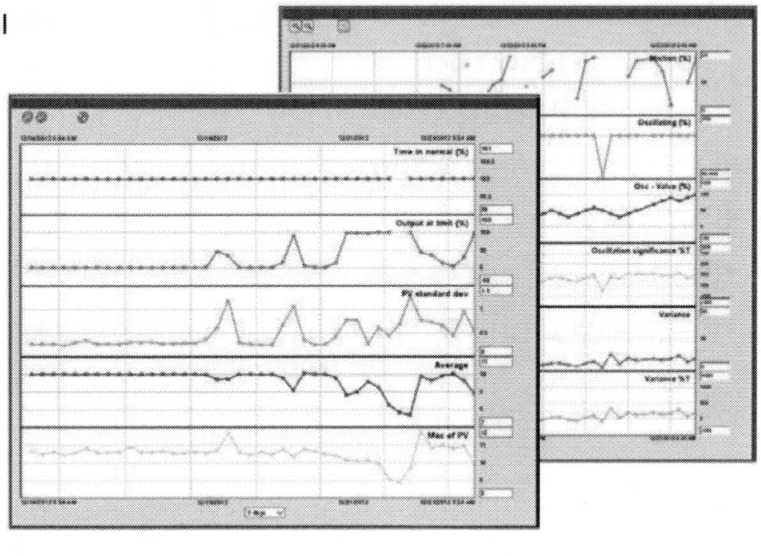

6

16AIChE Spring Meeting, Houston, TX
12TH GLOBAL CONGRESS ON PROCESS SAFETY

Automated Data to Information

Control Performance Monitoring

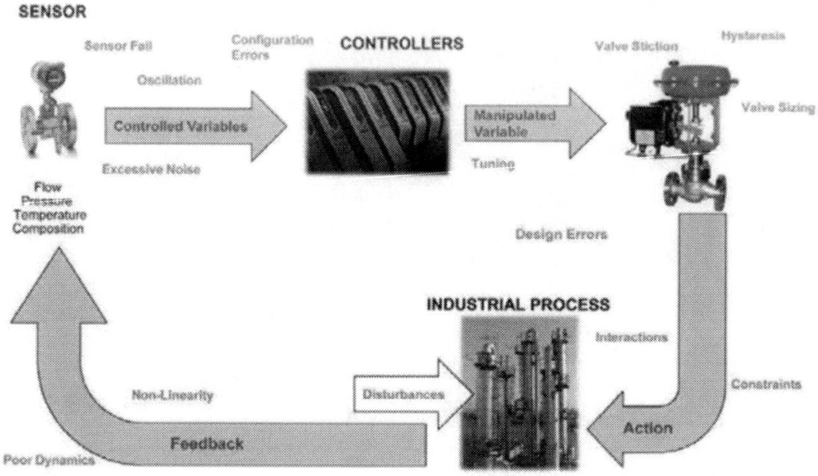

7

16AIChE Spring Meeting, Houston, TX
12TH GLOBAL CONGRESS ON PROCESS SAFETY

Example: Data to Information

Measuring Valve Stiction

- Is loop in AUTO?
- Is it Oscillating?
- Does it have a Valve?
- PV = Square Wave
- CO = Saw Tooth
- Peak-to-Trough of CO

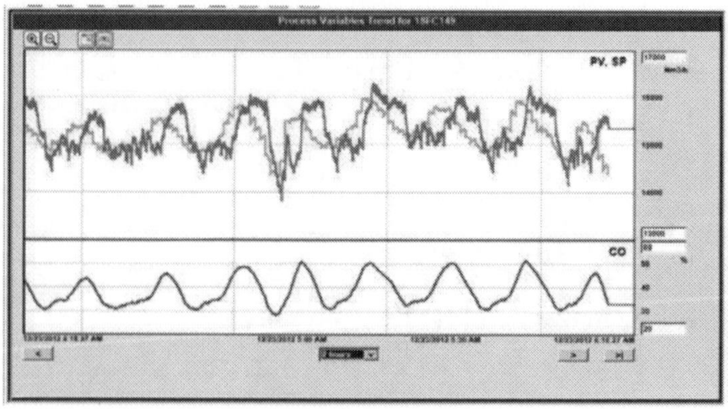

Existing Control Performance

- Control Performance – The Facts
  - 10%-35% of control loops are in **manual**
  - 1.5M to 5.3M in under utilized assets by running in manual in a 1000 loop site
  - 30% of control valves have problems
  - Are you repairing the correct valves?
  - 30% of control loops are tuned incorrectly, increasing variability in the process.
  - A tremendous amount of money can be saved by understanding the control loop interactions and implementing corrective action.

16AIChE
Spring Meeting, Houston, TX
12TH GLOBAL CONGRESS ON PROCESS SAFETY

Information to Action

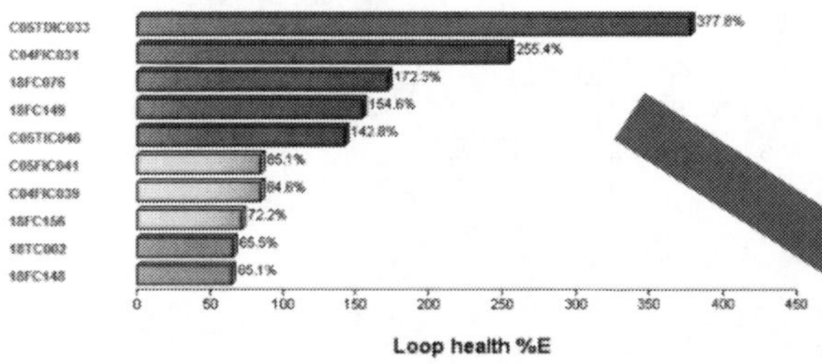

INFORMATION

ACTION

10

16AIChE Spring Meeting, Houston, TX
12TH GLOBAL CONGRESS ON PROCESS SAFETY

Analysis of a Typical Plant

One-Time Analysis to Identify Opportunity

260 Loops			
% Loops	Issue	Typical	World-Class
46	Loops Unable to Hold SP	1 - 5%	< 1%
89	Loops Not in Normal mode	10% - 25%	< 10%
35	Oscillating	10% - 40%	< 10%
40	Excessive Valve Travel	5% - 15%	< 5%
89	Valve or CO at Limits	5% - 15%	< 5%
53	Valve Stiction	10% - 20%	< 10%
15	Sluggish Tuning	3% - 5%	< 3%
3	Sensor Problems	1%- 3%	< 1%

11

16AIChE Spring Meeting, Houston, TX
12TH GLOBAL CONGRESS ON PROCESS SAFETY

Plenty of Data

Distillation column swings

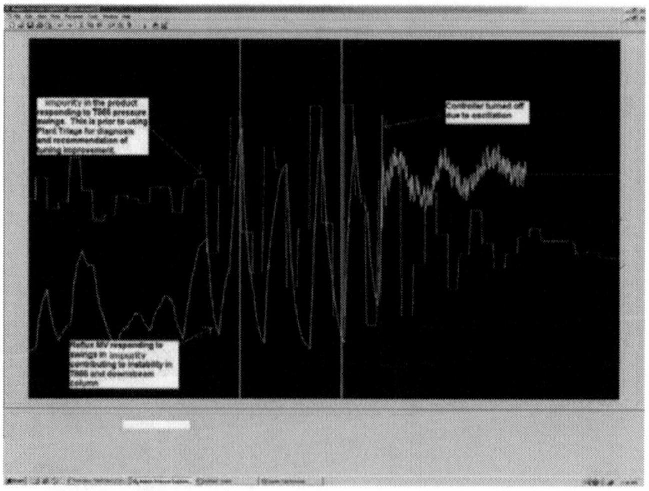

12

Use the Data to Obtain Information

Process mass amounts of data into a root cause

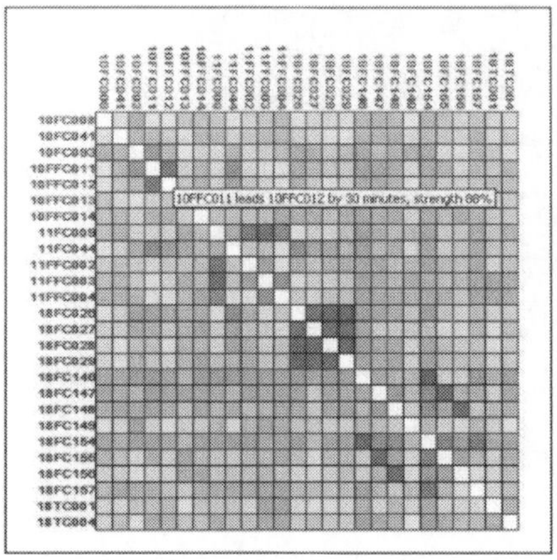

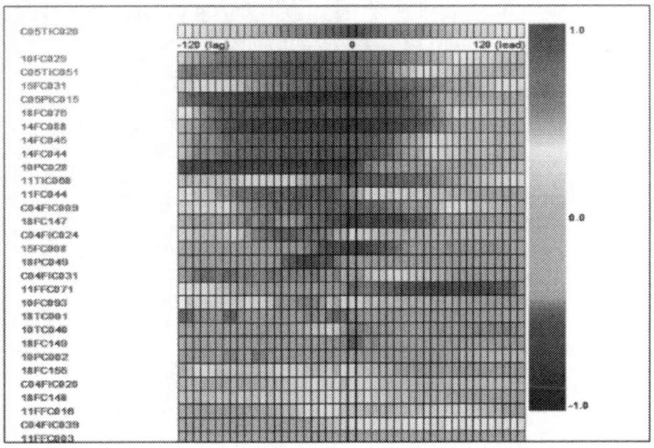

13

16AIChE Spring Meeting, Houston, TX
12TH GLOBAL CONGRESS ON PROCESS SAFETY

Take Action

Tune the Pressure controller -use "data" to obtain a process model

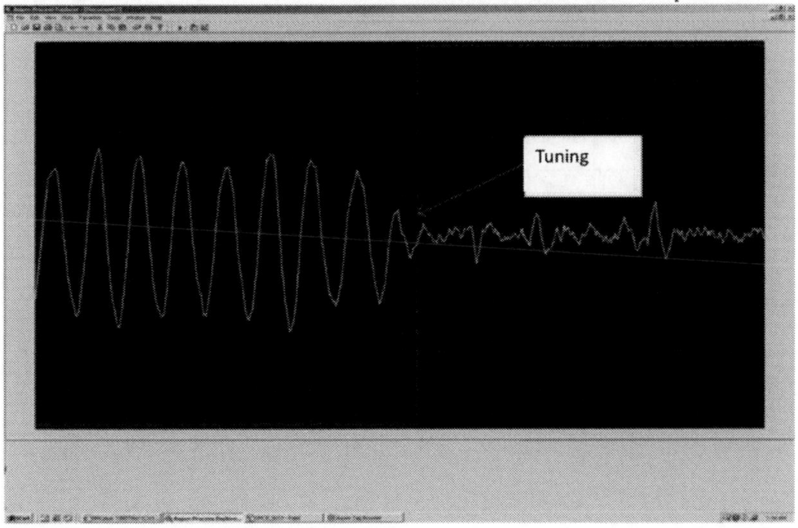

14

16AIChE Spring Meeting, Houston, TX

12TH GLOBAL CONGRESS ON PROCESS SAFETY

See Results

More stable and energy savings

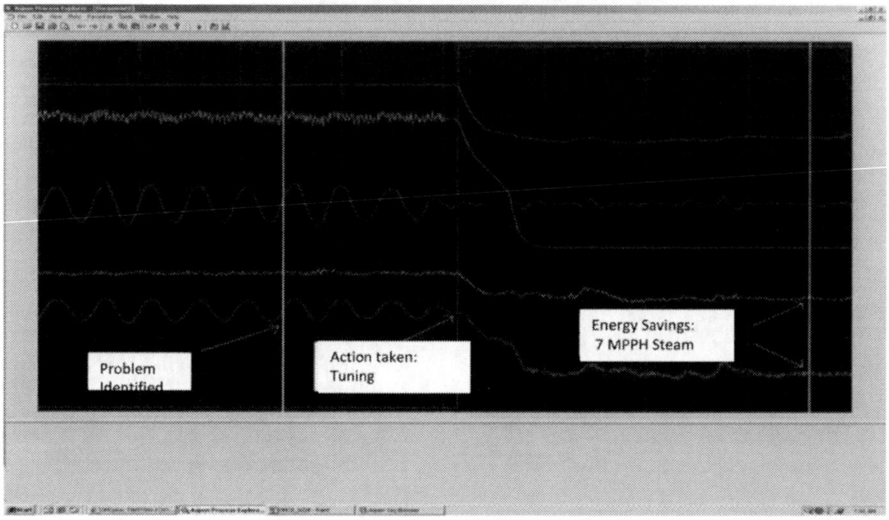

15

16AIChE Spring Meeting, Houston, TX

12TH GLOBAL CONGRESS ON PROCESS SAFETY

Example: Identify Tuning Opportunities

Active Model Capture

Models for 77 Loops

Loop	Description	Harris (normalized) (0-1)	Relative response time (sec)	Model quality	Process dead time (sec)	Process gain	Process lag 1 (sec)	Process lag 2 (sec)
Sort	Sort	Sort				Sort	Sort	Sort
38TC-6046	Темп.разбавитель	0.9982*	[illegible]	[illegible]	[illegible]	0.25†*	84†	0†
2220-LC-3790	L БАК 1ст.СОРТИРОВ./L TANK 1st SORTIROV	0.9874*	[illegible]	[illegible]	[illegible]	7.2†*	43.2†	0†
2210-TC-3621	Пар из конд.перв.пар/Pair of kond.perv.par	0.9865*	830†	3†	63.01†	6.8†*	482†	0†
2210-TC-3220	Темп. после нагрева/Temp. after heating	0.9859*	1008†	3†	61.29†	3.4†*	9000†	0†
2220-LV-3965	БАК ФИЛЬТРАТА 233.11/LHC LEACHATE 233.11	0.9847*	2†	1†	1†	1†	1E-08†	[illegible]
38FC-6127	Поток на входе пресса/Stream entering the press	0.9888*	210†	1†	13.28†	1.2†	0†	[illegible]
38FC-6168	Поток на входе пресса/Stream entering the press	0.9867*	290†	2†	35.93†	1.9†	0†	[illegible]
38FC-6087	Поток разбав.выдувн. бака/Feed razbav.vyduvn. tan	0.9856*	46†	3†	15.1†	0.54†	1.2†	[illegible]
2210-TCZ-3086	Пар НД в низ 6унк./OD pairs in the bottom of his e	0.9856*	620†	2†	96.26†	0.17†*	54.6†	[illegible]

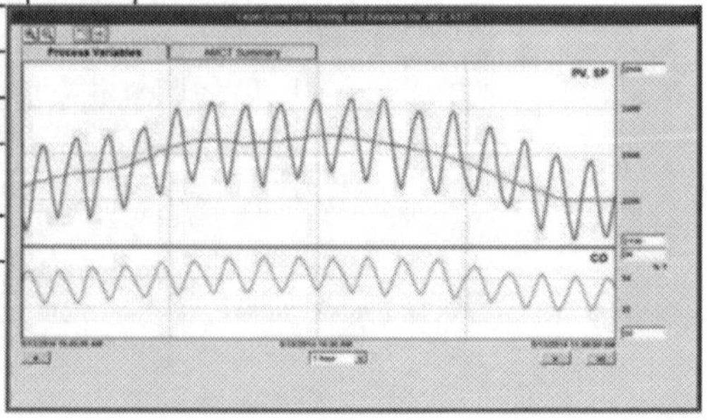

16

16AIChE Spring Meeting, Houston, TX

12TH GLOBAL CONGRESS ON PROCESS SAFETY

Practical Suggestions

An Approach to Control Performance

PLATFORM
- Software
- Hardware
- Installation
- Training
- Site Evaluation

SURGE
- Detailed Unit Evaluation
- Diagnosis
- Corrective Actions
- Value Documentation

SUSTAIN
- Continuous Improvement
- Diagnose & Correct New Issues
- Identify Additional Surges

17

16AIChE Spring Meeting, Houston, TX

12TH GLOBAL CONGRESS ON PROCESS SAFETY

More Practical Suggestions

- Do a Quick Survey
- Blanket an Area
- Don't be Overly Selective – i.e. Critical Loops Only
- Start with the Basics (Instrument, Valve, Tuning)
- Boost Data Collection Frequencies
- Follow Up is Key
- Learn from Others
- Document the Value

18

Questions & Answers

19

AUTHOR INDEX

AUTHOR INDEX

AUTHOR INDEX

AIChE
120 Wall Street, FL 23
New York, NY 10005-4020

CURRAN ASSOCIATES INC.
proceedings.com

ISBN 978-1-5108-2481-2
9 781510 824812